ON THE FACE STABILITY OF SHALLOW TUNNELS IN SAND

ADVANCES IN GEOTECHNICAL ENGINEERING AND TUNNELLING

16

General editor:

D. Kolymbas

University of Innsbruck, Division of Geotechnical and Tunnel Engineering

In the same series (A.A.BALKEMA):

1. D. Kolymbas (2000), *Introduction to hypoplasticity*, 104 pages, ISBN 90-5809-306-9
2. W. Fellin (2000), *Rütteldruckverdichtung als plastodynamisches Problem (Deep vibration compaction as a plastodynamic problem)*, 344 pages, ISBN 90-5809-315-8
3. D. Kolymbas & W. Fellin (2000), *Compaction of soils, granulates and powders - International workshop on compaction of soils, granulates, powders*, Innsbruck, 28-29 February 2000, 344 pages, ISBN 90-5809-318-2

In the same series (LOGOS):

4. C. Bliem (2001), *3D Finite Element Berechnungen im Tunnelbau (3D finite element calculations in tunnelling)*, 220 pages, ISBN 3-89722-750-9
5. D. Kolymbas, ed. (2001), *Tunnelling Mechanics, Eurosummerschool, Innsbruck, 2001*, 403 pages, ISBN 3-89722-873-4
6. M. Fiedler (2001), *Nichtlineare Berechnung von Plattenfundamenten (Nonlinear Analysis of Mat Foundations)*, 163 pages, ISBN 3-8325-0031-6
7. W. Fellin (2003), *Geotechnik - Lernen mit Beispielen*, 230 pages, ISBN 3-8325- 0147-9

8. D. Kolymbas, ed. (2003), *Rational Tunnelling, Summerschool, Innsbruck 2003*, 428 pages, ISBN 3-8325-0350-1

9. D. Kolymbas, ed. (2004), *Fractals in Geotechnical Engineering, Exploratory Workshop, Innsbruck, 2003*, 174 pages, ISBN 3-8325-0583-0

10. P. Tanseng (2006), *Implementation of Hypoplasticity for Fast Lagrangian Simulations*, 125 pages, ISBN 3-8325-1073-7.

11. A. Laudahn (2006), *An Approach to 1g Modelling in Geotechnical Engineering with Soiltron*, 197 pages, ISBN 3-8325-1072-9.

12. L. Prinz von Baden (2005), *Alpine Bauweisen und Gefahrenmanagement*, 212 pages, ISBN 3-8325-0935-6.

13. D. Kolymbas, A. Laudahn, eds. (2005), *Rational Tunnelling, 2nd Summerschool, Innsbruck 2005*, 291 pages, ISBN 3-8325-1012-5.

14. T. Weifner (2006), *Review and Extensions of Hypoplastic Equations*, 240 pages, ISBN 978-3-8325-1404-4.

15. M. Mähr (2006), *Ground movements induced by shield tunnelling in noncohesive soils*, 168 pages, ISBN 978-3-8325-1361-0.

On the face stability of shallow tunnels in sand

Ansgar Kirsch
University of Innsbruck, Division of Geotechnical and Tunnel Engineering

E-mail: Ansgar.Kirsch@gmx.de
Homepage: http://www.uibk.ac.at/geotechnik

The first three volumes have been published by Balkema
and can be ordered from:

A.A. Balkema Publishers
P.O.Box 1675
NL-3000 BR Rotterdam
e-mail: orders@swets.nl
website: www.balkema.nl

Bibliographic information published by the Deutsche Nationalbibliothek

The Deutsche Nationalbibliothek lists this publication in the Deutsche Nationalbibliografie; detailed bibliographic data are available in the Internet at http://dnb.d-nb.de.

ISBN 978-3-8325-2149-3
ISSN 1566-6182

Logos Verlag Berlin GmbH
Comeniushof, Gubener Str. 47,
10243 Berlin
Tel.: +49 030 42 85 10 90
Fax: +49 030 42 85 10 92
INTERNET: http://www.logos-verlag.de

It's not the things you don't know that get you into trouble. It is the things you think you know for sure.

(attributed to Sir Winston Churchill)

Preface by the author

During construction of shallow tunnels the face stability is an important issue. To minimise settlements at the ground surface and to prevent an uncontrolled collapse of the soil ahead of the tunnel a *necessary support pressure* must be calculated. A comparison of various models, which were proposed for this purpose, revealed a large amount of scatter of the different predictions. Still, only a few models have found their way into engineering practice. So do we know for sure that the tunnel faces will be stable?

The large discrepancy of model predictions served as starting point for this work. Existing theoretical and numerical models were compared on the basis of small-scale laboratory experiments and numerical simulations.

The results of the performed investigation indicate that most of the theoretical models in use make conservative predictions. Only two of the models in question were able to predict the experimental results on a 95% confidence level. Finite element simulations of the performed experiments were able to reproduce both, the observed displacements and necessary support pressures.

The topic of this research has been tackled in many ways. I hope that my results may serve as reference for those who apply face stability models in engineering practice, but also for those who develop new models and need experimental evidence.

I am deeply grateful to my advisor Prof. Dimitrios Kolymbas for his steady support and invaluable guidance throughout this research work. He was *always* available for discussion and provided many stimulating thoughts.

Also the advice by Prof. David Muir Wood, who added with his valuable remarks to the quality of this work, is very much appreciated.

My research work would not have been possible without various kinds of support, for which I am deeply thankful:

- *scientific* support by my colleagues and friends at the University of Innsbruck and in the geotechnical "world",
- *technical* and *administrative* support by the staff of the institute and my former diploma students,
- *financial* support by the Tyrolean Science Foundation,

- *conceptual* support by the German National Academic Foundation.

I very much appreciated the open-minded and motivating atmosphere at the institute.

Finally, I am grateful to my family for their unconditional *moral* support throughout all the time at university. Love and affection go to Lavinia and Emma: you make my days brighter.

To Lavinia, the one who chose me.

Contents

Chapter 1

Introduction

The number of tunnelling projects in loose ground has increased significantly in the last couple of years. Especially in densely populated areas, the demand for quick transport of people and goods has led to continuously growing underground systems. To limit tunnel lengths and to allow for easy access, urban tunnels are designed as close to the ground surface as possible. Tunnels with covers of only a few metres have been constructed successfully.

Shield heading

Tunnels with low covers are often headed using the shield technique. It can prevent excessive ground movements during tunnel advance and secures a safe working environment for the support works. The shield, which is basically a large diameter steel tube, is pushed forward into the ground with hydraulic jacks that use the already installed lining as abutment.

(a) Rotary cutter head, *Herrenknecht AG*

(b) Roadheader, *Herrenknecht AG*

Figure 1.1: Excavation tools for shield machines

Under cover of the shield the ground is excavated full-faced, with rotary cutter heads (Fig. 1.1 a), or partly, with roadheaders (Fig. 1.1 b). In the tail of the shield, which has an approximate length of 0.8 times the tunnel diameter D, the segmental lining is installed and the remaining tail gap grouted with cement. Fig. 1.2 illustrates the sequences of shield heading.

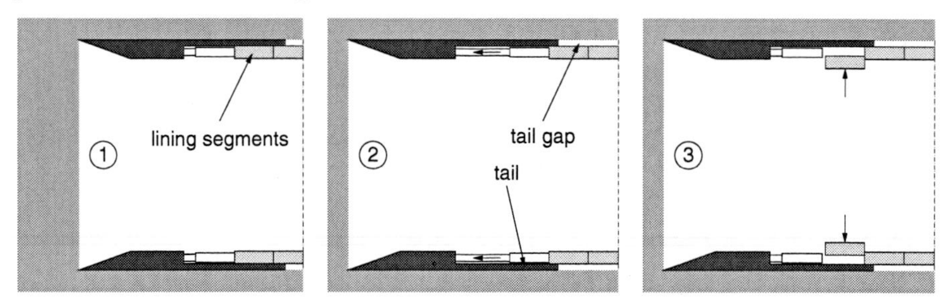

Figure 1.2: Sequences of shield heading, [73]

Shield heading is cost-effective for lengthy tunnels. Under favourable ground conditions high advance rates can be achieved.

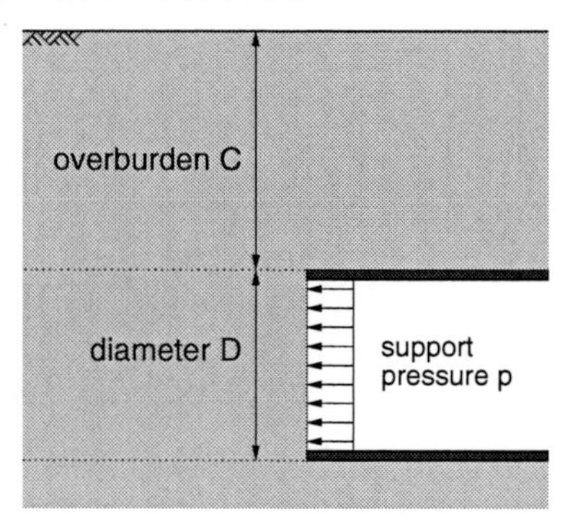

Figure 1.3: Tunnel geometry

Fig. 1.3 illustrates the geometrical variables used in the following.

Face support

To minimise settlements at the ground surface and to prevent inrush of the soil ahead of the tunnel, the face must be supported. Various techniques have evolved which will be explained briefly; for a detailed description see e.g. *Maidl* et al. [82].

When the tunnel is driven above the groundwater table, there are two major possibilities for face stabilisation:

- **Face panels** support the excavated surface (Fig. 1.4 a). Advance is achieved step by step by opening "windows" on the tunnel face and subsequently excavating small sections. Those openings are temporarily supported with e.g. shotcrete.

(a) Face panels, *Wayss & Freytag*

(b) Closed cutter head, *World Tunnelling, 12/2001*

Figure 1.4: Face stabilisation methods above the groundwater table

- A **closed cutter head** supports the tunnel face over its entire area (Fig. 1.4 b). The soil enters the excavation chamber via narrow slots in the cutter head. This type of support can be applied in temporarily stable ground with low permeability [82].

Matters become more complicated when the tunnel is driven below the groundwater table or when maintenance works have to be done ahead of the shield.

- The use of **compressed air** in the excavation chamber prevents groundwater ingress. Either the air in the whole tunnel or only in the front part of the shield is pressurised. Additional methods need to be applied, though, to counteract the earth pressure (e.g. above mentioned ones).

  Restrictions on the use of this method mainly arise from the allowable air pressure regime for workers.[1] Equipment and workers are, moreover, endangered by blow-outs of the air which lead to a sudden drop in pressure head and may result in a inrush of water and soil. The advance speed with this method is lower than under atmospheric conditions because of necessary time spans for decompression of the personnel in the air lock.

- **Slurry shield**[2] machines (Fig. 1.5 a) use a pressurised chamber to stabilise the face. Thus, the remaining working environment remains under atmospheric pressure. A bentonite slurry inside the pressure chamber penetrates into the ground and forms a mud cake over which the necessary support pressure can

[1]The maximum pressure is limited for health reasons; legally allowable pressures reach from $\approx$ 3.6 bar (e.g. Germany, United Kingdom) up to 4.5 bar (The Netherlands) or 4.8 bar (France). High pressures endanger the workers' health by nitrogen narcosis [127].

[2]Also specific trade names, such as *Hydroshield*, *Hydrojetshield* and *Thixshield*, refer to slurry shields [82].

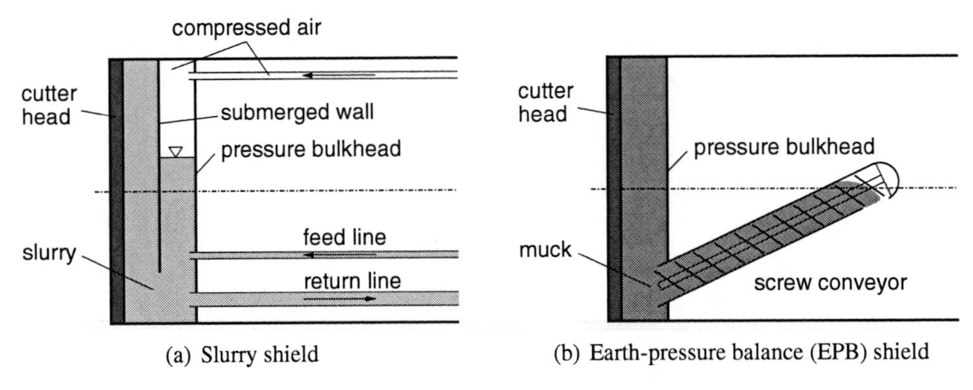

(a) Slurry shield (b) Earth-pressure balance (EPB) shield

Figure 1.5: Schematic diagram of shield machines

be applied onto the soil. To maintain and fine-tune the prescribed pressure, an air reservoir is used as part of the pressure chamber.[3]

The soil is excavated with a cutter head and the soil-bentonite mixture is pumped away to a separation plant where the bentonite suspension is recycled. The separation process only works well in cohesionless soil that contains little or no clay and silt, which puts a limit to the applicability of slurry shields.

- **Earth-pressure balance (EPB) shields** (Fig. 1.5 b) are an alternative to slurry shields in weak cohesive soils. The face is supported with a mud, which consists of the excavated soil, water and additives (e.g. polymer foam).

  The muck in the EPB shield is pressurised to counteract the earth pressure. The pressure in the chamber is controlled by coordinating the advance speed of the shield and the quantity of the removed muck. As this procedure is not very precise, the pressure can only be controlled within ± 0.5 bar [73].

- **Combined shields**[4] have been developed to cope with varying ground conditions, even with geological strata of soft soil and hard rock on the same drive-operation. In simple terms, slurry shield machines are equipped with additional cutting tools for hard rock and special outlets for coarse material. Or, EPB shield machines are modified in a way that the excavation chamber can be filled with pressurised slurry.

For a long time the applicability of the different advance methods depended mainly on the grain size distribution of the encountered soil (Fig. 1.6). Nowadays, with proper conditioning agents EPB shields can very well be used in sand, whereas

[3]As air is much more compressible than the slurry, the pressure in the air cushion is much less sensitive to volume losses. The pressure in the air phase can be controlled within ± 0.1 bar.

[4]Trade names for combined shields are e.g. *Polyshield* or *Mixshield* [82].

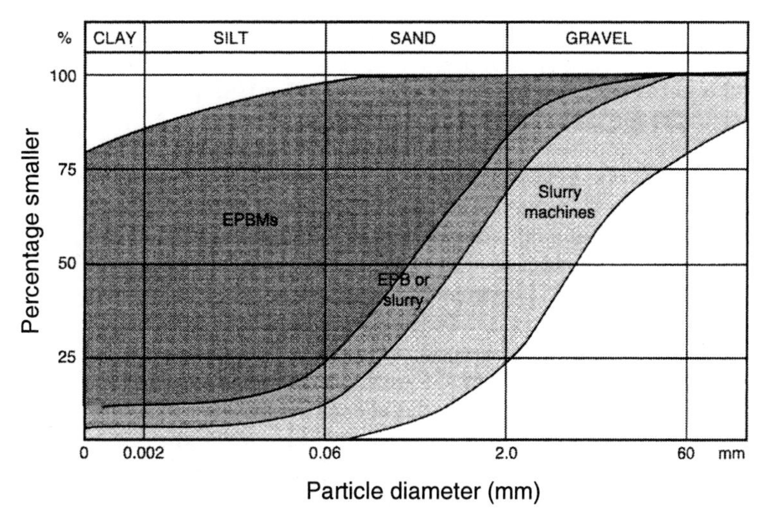

Figure 1.6: Range of applicability of slurry and EPB shields in terms of grain size distributions, *British Tunnelling Society* [126]

slurry shields can be applied in clayey soils [127]. Other factors, such as stability of the tunnel face, potential to reduce surface settlements, ground water control, tool wear, advance speed and operational as well as investment costs have become equally important.

Face stability analysis

The estimation of the required support pressures for slurry- or earth-pressure balance shields to ensure global stability of the face has been a topic of research until the present day.

For a slurry shield, three stages of face support have to be considered:

- excavation chamber completely filled with slurry (standard mode of operation, Fig. 1.7 a),
- partial drawdown of slurry (Fig. 1.7 b),
- complete drawdown of slurry (Fig. 1.7 c).

Partial and complete drawdown are necessary for maintenance and repair works, or to recover boulders from the excavation chamber. In these cases, the upper part or the whole tunnel face is supported with pressurised air. Fig. 1.7 also shows the assumed pressure distribution in the ground and in the excavation chamber.

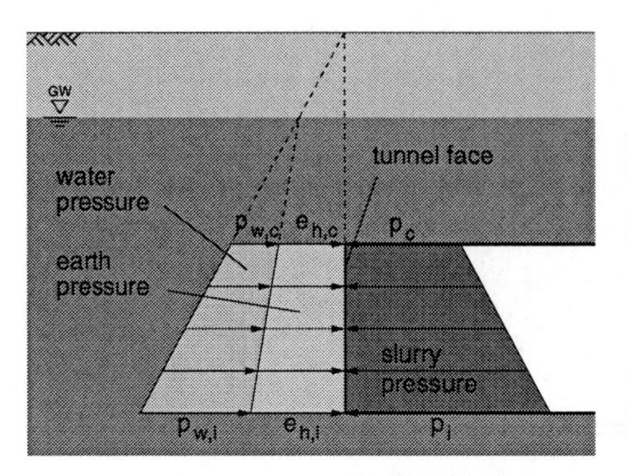

(a) Full support with slurry (standard)

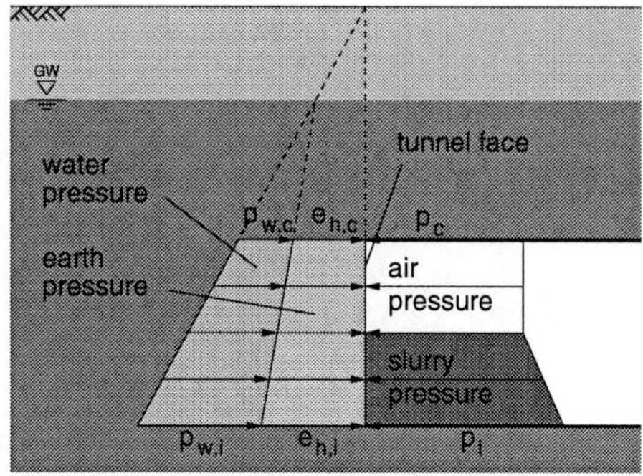

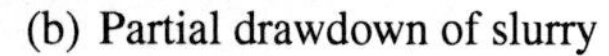

(b) Partial drawdown of slurry

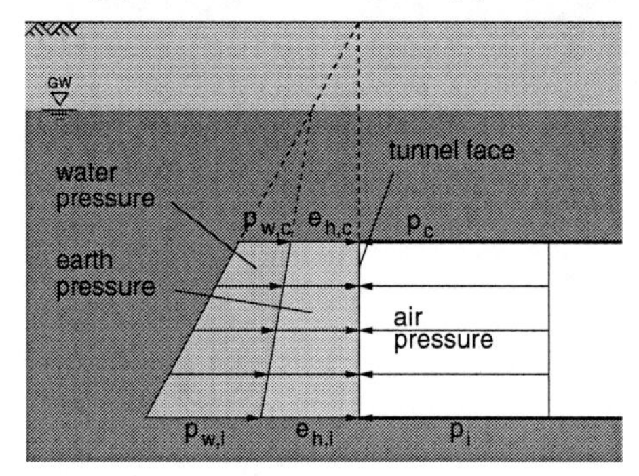

(c) Complete drawdown of slurry

Figure 1.7: Stages of face support for a slurry shield

The necessary support *force* S is composed of a resultant force from water pressure, W, and from effective earth pressure E. With corresponding (partial) safety factors, η_E and η_W, the following relation for S is generally adopted for overall stability[5,6]

$$S \geq \eta_E \, E + \eta_W \, W \quad . \tag{1.1}$$

The partial safety factors η_E and η_W are not explicitly given in codes of practice; $\eta_W = 1.0 \ldots 1.05$ and $\eta_E = 1.10 \ldots 1.75$ can be found in the literature (e.g. *STUVA* [118], *ZTV-Ing.* [21], *Balthaus* [12], *Jancsecz* et al. [58]). The large span for η_E indicates the uncertainty involved in the determination of the earth pressure on the tunnel face.

Apart from global safety for the three stages of support, also the pressure values at the tunnel crown, p_c, and invert, p_i, must be checked (variables in Fig. 1.7):

$$\begin{aligned} p_c &\geq \eta_c \, (e_{h,c} + p_{w,c}) \quad , \\ p_i &\geq p_{w,i} + \eta_i \quad . \end{aligned}$$

[5]Please note that the term *stability* is not used in its pure mechanical sense here; in a geotechnical engineering context it rather means a state of *structural safety*.

[6]*Internal* stability, i.e. the safety against fallout of single grains or grain groups, is not considered here.

Safety factors η_c = 1.1 and η_i = 10 ... 20 kPa have been recommended by, e.g., *STUVA* [118] or *Gabener* et al. [42].

The increase of slurry pressure with depth is hydrostatic:

$$p_i = p_c + D\,\gamma_{\text{slurry}} \quad ,$$

with γ_{slurry} = 10.5 kN/m^2 for standstill and γ_{slurry} = 12.0 kN/m^2 for advance.[7]

The integral of pressure p over the tunnel face, i.e. the resultant force S, must fulfil (1.1) for all modes of operation.

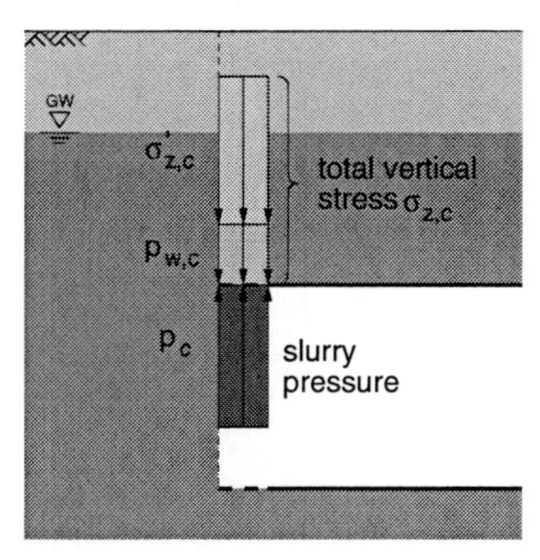

Figure 1.8: Assumed stresses for safety analysis against upheaval of the overlying ground

An additional aspect of face stability analysis is the check of safety against blow-outs or upheaval of the overlying ground (Fig. 1.8). Different models have been suggested (cf. *Broere* [19]); two commonly used versions are

$$\begin{aligned} \sigma'_{z,c} + p_{w,c} &\geq \eta\, p_c \qquad \text{or} \\ \sigma'_{z,c} &\geq \eta\,(p_c - p_{w,c}) \quad . \end{aligned}$$

Neglecting the shear strength of the soil ($\tau_f = 0$), $\sigma'_{z,c}$ is usually calculated as product of effective self-weight γ' and overburden C. Values for η between 1.0 and 1.2 were suggested (e.g. *STUVA* [118], *ZTV-Ing.* [21]).

[7] During advance the slurry carries a larger portion of soil particles that increase the self-weight of the slurry.

Motivation for and outline of this work

This short overview raises one main problem of face stability analysis: while the water pressure can be predicted reasonably well, as indicated by the low η_W, the determination of the resultant earth pressure on the face is rather vague.

Economic implications for the design of a tunnel arise from the safety against blow-outs: a larger design earth pressure, $\eta_E \cdot e_{h,c}$, leads to a larger necessary overburden to provide safety and, consequently, a deeper and longer tunnel. Thus, safety, construction and maintenance costs depend on the design earth pressure [8].

Also technical issues favour support pressures as low as possible: *Broere* [19] mentioned "the tendency to work with a low support pressure in order to minimise friction and maximise excavation speed" during tunnel heading in soft ground.

Project	**year**	***D* (m)**	**Failure**
Habsburgtunnel, *Switzerland*	1988-1994	11.5	two daylight collapses, i.e. cave-in of the material ahead of the tunnel face
Grauholztunnel, *Switzerland*	1990-1992	10.6	daylight collapse, local collapses
Fäsenstaubtunnel, *Switzerland*	1993	11.4	collapse of a weak sand layer above the tunnel crown
Oenzbergtunnel, *Switzerland*	1994-2000	12.4	daylight collapses
Weser tunnel, *Germany*	1999	11.7	local collapse, daylight collapse, [80]
Metro tunnel Porto, *Portugal*	2000	8.9	3 collapses, [43]

Table 1.1: Examples for failure events in shield driven tunnels [115]

In spite of all progress in research and technology, face collapses during construction of shallow tunnels still occur (Tab. 1.1). High additional costs for stabilisation works and recovery of machine or tools arise. Also, construction times extend significantly. Reports of failure incidents usually remain unpublished because the involved parties fear a damage of their reputation. In some cases, the publication of data is also prohibited by long-lasting legal procedures.

But how can failure incidents be reduced?

The question of face stability has been tackled in different ways. **Chapter 2** of this work gives an overview of existing models and approaches to tackle the problem of face stability, with a focus on shallow tunnels. Some representative approaches are compared on the basis of a simple example, and the sensitivity of the predicted support pressures with respect to the input parameters is worked out. The chapter finishes with an overview of current engineering practice concerning face stability calculations. It will be shown that a satisfactory understanding of the mechanisms together with a reliable way for calculating the support pressure is still missing.

To further assess the quality of proposed models for face stability analysis, the author performed two series of small-scale model experiments, which are described in **Chapter 3**. Scaling laws are indispensable for the interpretation of small-scale experiments; therefore, they are shortly introduced.

In the first series of experiments the evolution of failure mechanisms in dense and loose sand with different overburden were investigated, making use of Particle Image Velocimetry. The resulting support force on the tunnel face was studied in a second series of experiments. An overview of the properties of sand at low stress levels around 2 kPa together with a comparison of the author's experimental results with predictions by several theoretical/numerical models concludes the chapter.

Chapter 4 concentrates on the investigation of face stability with finite elements. Finite element calculations are increasingly applied in engineering practice, but the choice of material model, input parameters and numerical setup is not straightforward. The author assessed two constitutive models with respect to their ability to reproduce his experimental observations.

Chapter 5 finally puts the findings together to provide recommendations for the application of existing numerical and theoretical models for the prediction of necessary support pressures for a tunnel face.

SYNOPSIS

The face stability of shallow tunnels must be guaranteed to minimise settlements at the ground surface and to prevent collapse of the soil ahead of the tunnel. For slurry and EPB shield machines a *necessary support pressure* must be prescribed to counteract water and earth pressure with a sufficient safety margin. Proposed partial safety factors on the earth pressure are relatively high, indicating a certain amount of uncertainty in its determination.

The purpose of this study was to recommend appropriate theoretical and numerical models for the given problem. Therefore, existing theoretical and numerical models were compared to results of the author's small-scale laboratory experiments and numerical simulations.

Chapter 2

State of the art of tunnel face stability analysis

This chapter compiles different approaches to analyse the face stability of shallow tunnels in (mainly) cohesionless material. Theoretical derivations for the necessary support pressure as well as laboratory tests and numerical calculations are covered. An example calculation illustrates the disagreement of some chosen models in comparison with each other and to experimental/numerical observations. A sensitivity study shows how these models react to a variation of input parameters. The current engineering practice and some examples conclude the chapter.

2.1 Theoretical models

Various theoretical models to predict the necessary support pressure for tunnels in soft ground have been put forward. These approaches can be subdivided into kinematic approaches with failure mechanisms and static approaches with admissible stress fields. Some additional approaches are neither kinematic nor static.

2.1.1 Kinematic approaches

Horn [55] was the first to present a translational rigid block failure mechanism for the given problem. As his model has been used and extended in many subsequent works, and is frequently used in engineering practice, it will be presented in detail.

The *Horn* model assumes a sliding wedge mechanism, consisting of a chimney and a prismatic wedge (Fig. 2.1 a). The tunnel cross section is approximated by a rectangle (or square) with the same area. Fig. 2.1 b shows the forces which are acting on the sliding wedge:

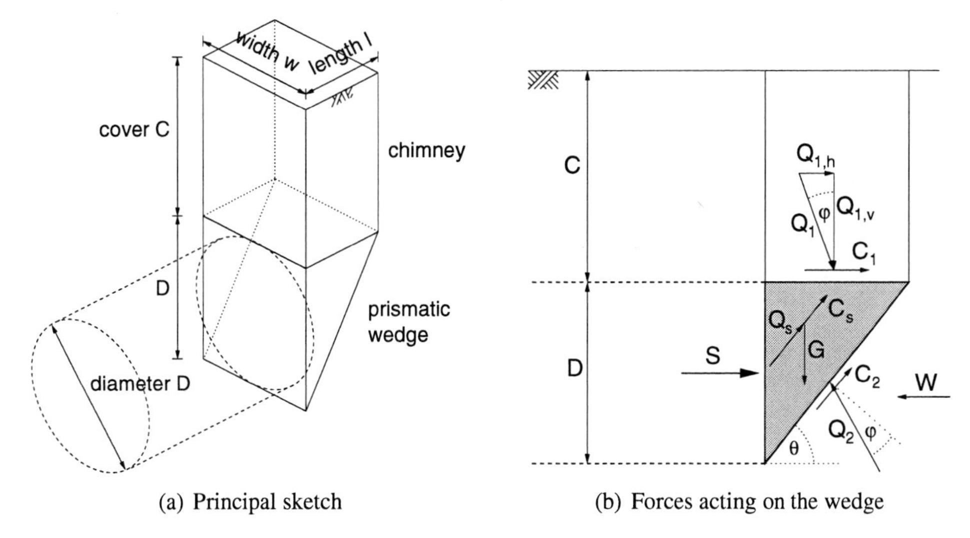

(a) Principal sketch (b) Forces acting on the wedge

Figure 2.1: Sliding wedge mechanism, proposed by *Horn*

- On top of the wedge acts a vertical force $Q_{1,v}$, which is in most cases calculated with the silo equation. *Janssen* [61] presented this theory for the distribution of vertical stress in a silo. A derivation of the silo equation can be found e.g. in *Kirsch* and *Kolymbas* [68] or *Ruse* [107].
  In addition, *Terzaghi* [125] put forward the idea of a minimum load Q_v resulting from the self-weight of a *keystone* above the wedge.

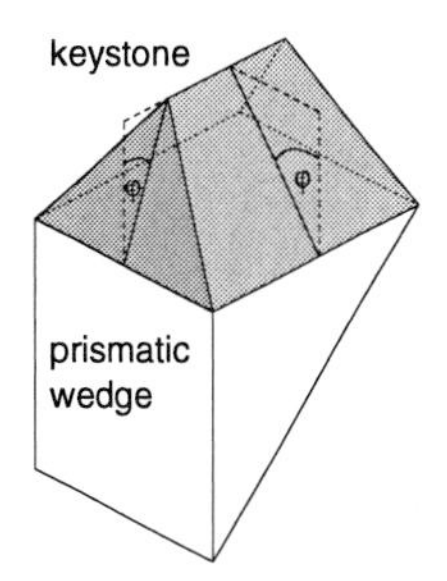

- Interaction forces, such as a horizontal cohesion force C_1 and a frictional force $Q_{1,h}$ on top of the wedge, and a cohesion force C_2 and a reaction force Q_2 on the inclined slip surface, must be taken into account.
- The force G is the self-weight of the sliding wedge.
- In case of water-saturation, W accounts for water pressure.

For the derivation of the **silo equation**, a disc of infinitesimal height dz is cut out of the silo body. By equilibrating the vertical forces acting on the disc, the distribution of vertical stress $\sigma_v(z)$ in a silo can be calculated:

$$\sigma_v(z) = \frac{\gamma A - c\,U}{U\,K_{\text{silo}}\tan\varphi}\left[1 - \exp\left(K_{\text{silo}}\frac{U}{A}\tan\varphi z\right)\right]$$

In this equation γ is the self-weight of the soil, K the earth pressure coefficient, φ the friction angle and c the cohesion. The geometry is taken into account in terms of area A and circumference U of the disc. The vertical stress $\sigma_v(z)$ is assumed constant over the area of the disk.

- Finally, cohesion and frictional forces C_s and Q_s act on each vertical triangular face of the sliding wedge.

Via horizontal and vertical force equilibrium, the support force can be calculated as a function of ϑ, the inclination of the basal surface of the wedge. The *necessary* support force S results from maximisation: $S = \max S(\vartheta)$ for a range of ϑ between 0° and 90°. In some cases, ϑ is assumed to be $45 + \varphi/2$ (e.g. *Katzenbach* and *Strüber* [65]).

Unfortunately, the proposed model offers quite a few configuration options to the user. These are briefly mentioned with suggestions from the literature:

- Lateral earth pressure coefficients for the silo equation:
  - $K_{\text{silo}} = 0.8$ (*Anagnostou* and *Kovári* [3, 4]),
  - $K_{\text{silo}} = 1.0$ (*Mayer* et al. [86]),
  - $K_{\text{silo}} = 0.7 \ldots 1.5$ (*Girmscheid* [45]),
  - $K_{\text{silo}} = K_0 = 1 - \sin\varphi$ (*Kirsch* and *Kolymbas* [68]).[1]
- Lateral earth pressure coefficients for the calculation of the horizontal pressure on the wedge:
  - $K_{\text{wedge}} = 0.4$ (*Anagnostou* and *Kovári* [3, 4]),
  - $K_{\text{wedge}} = 0.4 \ldots 0.5$ (*Girmscheid* [45]),
  - $K_{\text{wedge}} = 0$; on the safe side frictional forces are neglected (*Mayer* et al. [86]),
  - $K_{\text{wedge}} = K_0 = 1 - \sin\varphi$ (*Kirsch* and *Kolymbas* [68], *DIN 4126* [35]).

  The proper choice of both earth pressure coefficients is difficult. In the author's opinion, a parameter/sensitivity study for K_{silo} and K_{wedge} should be performed to assess their influence on the resulting necessary support force.
- Distribution (with depth) of vertical and, thus, horizontal earth pressure acting on the sides of the wedge:
  - linear increase with depth from the ground surface (Fig. 2.2 a, *Kirsch* and *Kolymbas* [68])
  - pressure on the crown level of the tunnel according to silo theory; linear increase with depth from there downwards (Fig. 2.2 b, *Girmscheid* [45])

[1] $K_0 = 1 - \sin\varphi$ is *Jaky*'s approximation for the earth pressure at rest of un-preloaded cohesionless material.

- pressure on the crown level of the tunnel according to silo theory; from there downwards according to German Standard for diaphragm walls *DIN4126* [35], i.e. linear increase up to an additional depth equal to width w; from there on constant with depth (Fig. 2.2 c, *Anagnostou* and *Kovári* [3, 4]).

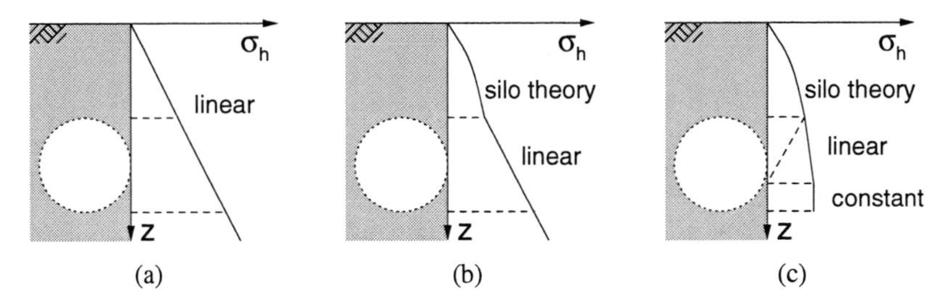

Figure 2.2: Possible earth pressure distributions for the *Horn* model

- Consideration of cohesion and horizontal frictional force on top of the wedge:
  - both C_1 and $Q_{1,h}$ (*Kirsch* and *Kolymbas* [68]),
  - neither C_1 nor $Q_{1,h}$ (*Anagnostou* and *Kovári* [3, 4], *Mayer* et al. [86], *Holzhäuser* [53]).

  The neglection of C_1 and $Q_{1,h}$ is an arbitrary modification of the model and violates the underlying mechanics: if two blocks slide along each other, there must be interaction forces.

- Approximation of the mostly circular tunnel cross section:
  - with a square (*Anagnostou* and *Kovári* [3, 4], *Mayer* et al. [86], *Girmscheid* [45]),
  - with a rectangle of height D (*Holzhäuser* [53], *Maidl* [83]).

  In most cases square and rectangle are assumed to have the same cross sectional area as the tunnel. But, the choice of a square leads to a change in geometry of the problem: either the overburden height must be changed or the location of the centre of area of the face must be moved. These problems can be avoided with a rectangle or a square with side lengths D.

- Shape of the basal boundary of the sliding wedge:
  - straight line (*Anagnostou* and *Kovári* [4], *Kirsch* and *Kolymbas* [68], *Girmscheid* [45], *Mayer* et al. [86]),
  - circular arc (*Krause* [75]),
  - logarithmic spiral (*Maidl* et al. [82], *Mohkam* and *Wong* [90], *Murayama* cited in [75]).

A circular arc is only kinematically possible, if the material is assumed to shear without volumetric strain. A logarithmic spiral is kinematically possible, if the material is assumed to dilate during shear, e.g. making use of an associated flow rule. This assumption is rather unrealistic for sands at large shear strains.

This short overview shows that the *Horn* model leaves space for interpretation and modification, which will be further illustrated in an example calculation (Sec. 2.5).

Kirsch and *Kolymbas* [68] have indicated how to account for different soil layers *above* or *below* the tunnel. For inhomogeneous conditions *ahead* of the tunnel *Jancsecz* et al. [58] suggested to calculate average values for self-weight γ and friction angle φ, separately for the soil above the tunnel and ahead of the face. *Broere* [19] pointed out, though, that calculation with average properties for the soil in the sliding plane of the wedge leads to unrealistic results. Based on the *Horn* mechanism, solutions have been developed to consider strata with different material properties ahead of the tunnel face (e.g. *Broere* [19] or *Anagnostou* and *Serafeimidis* [5]).

Jancsecz and ***Steiner*** [60] presented a calculation method based on the *Horn* model, which has been used successfully in practice, according to the authors. From various example calculations, *Jancsecz* and *Steiner* derived a "three-dimensional earth pressure coefficient K_{a3}" [60] as function of cover-to-diameter ratio C/D and friction angle φ. This coefficient can be used as in common earth pressure calculations.

Unfortunately, the authors did not mention their configuration of the sliding wedge mechanism (especially earth pressure coefficients for the silo and on the sides of the wedge, and earth pressure distribution with depth). Moreover, there are errors in their force equilibrium equations. This makes it rather difficult to reproduce their results.

Other, more practically oriented aspects of tunnelling in soft ground have been studied with the *Horn* mechanism:

- Some publications (*Broere* [19], *Balthaus* [12], *Anagnostou* and *Kovári* [3], *Jancsecz* and *Steiner* [60], *Mohkam* and *Wong* [90]) focused on the interrelation between penetration depth of the slurry into the soil and applicable support pressure.

- *Anagnostou* and *Kovári* [3] presented theoretical results on the applicable support pressure, resp. support force, as a function of time for slurry shields.

- *Anagnostou* and *Kovári* [4] investigated the influence of water seepage into the excavation chamber of an EPB shield on the stability of the face.

- *Hochgürtel* [52] studied the face stability of tunnels with pressurised air support and the loss of pressure due to percolation of air. Also *Holzhäuser* [53] and *Jancsecz* and *Steiner* [60] investigated these aspects.

Vermeer et al. [131] proposed a three-dimensional sliding wedge mechanism. Similar to *Horn*'s, it consists of a soil block overlying a prismatic wedge. The authors introduced two failure modes (Fig. 2.3):

- For a *chimney-type failure* (Fig. 2.3 a) the resultant forces from the rigid soil block act as surcharge on the sliding wedge. In this case the two blocks remain in contact during failure, and the collapse mechanism reaches up to the ground surface.
- The *local failure* is visualised in Fig. 2.3 b. If the cohesion and frictional forces on the side of the block are large enough to keep it in place, only the wedge slides into the tunnel. This is the case [131] for

$$c \geq \frac{\gamma}{2} \left[\frac{wD \cot\vartheta}{w + D \cot\vartheta} - C \tan\varphi \right] \quad .$$

  The variables are chosen according to Fig. 2.1, w is the width of the wedge.

In both cases the necessary support force S is calculated via horizontal and vertical force equilibrium.

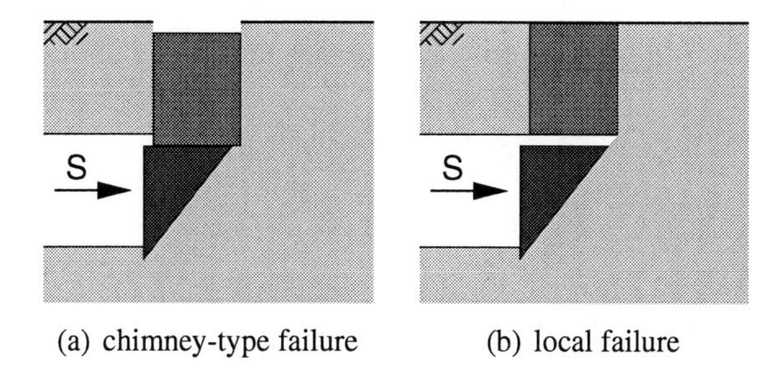

(a) chimney-type failure (b) local failure

Figure 2.3: Failure mechanisms proposed by *Vermeer* et al., [131]

The approach by *Vermeer* et al. has the drawback that the overlying soil is treated as a rigid block. Especially for non-cohesive soils the local failure seems rather unrealistic as soil particles would successively fill the gap between block and wedge.

Krause [75] considered a single hemispherical body of soil as three-dimensional failure mechanism. Cohesion and frictional forces in the contact area build up a resistance against a rotational type of failure around the y-axis through point 0 (Fig. 2.4). By equilibrating the moments acting on the body, he derived an expression for the necessary support pressure p_f.

Though the evolving formula is simple, the underlying mechanism is rather speculative. Given a support pressure p that is constant with depth over the whole tunnel face, a rotation of a hemispherical soil body without any movement above the tunnel

Within plasticity theory, the limit theorems, also called **bound theorems**, allow to estimate the collapse load F_{collapse}. For the **upper bound theorem** a *kinematically possible* collapse mechanism is chosen. If the dissipative work, needed to overcome the mechanism's shearing resistance, is smaller than the work done by body and surface forces, the body will collapse. For the face stability problem the upper bound theorem yields a support force that is lower or equal to F_{collapse}.
The accuracy of an upper bound solution depends on the similarity between real and chosen failure mechanism. A good introduction into bound theorems and their application to soil mechanics is given by *Chen* and *Liu* [27].

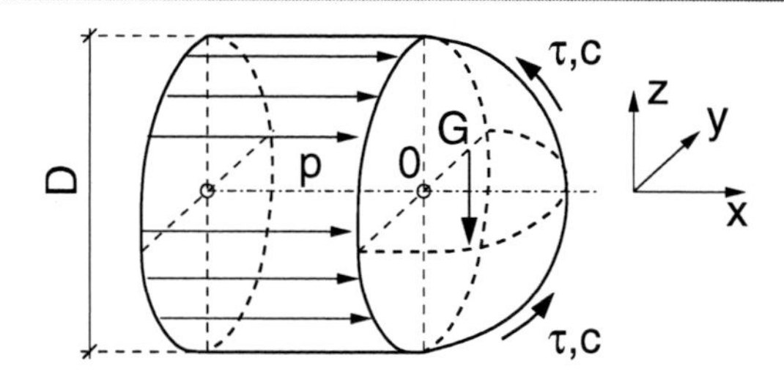

Figure 2.4: Failure mechanism by *Krause*, [75]

seems unlikely. Moreover, the horizontal force equilibrium on the soil body is *not* fulfilled.

Léca and ***Dormieux*** [77] presented a sliding wedge mechanism of one or two solid conical wedges with circular cross sections (Fig. 2.5). They calculated the necessary support pressure making use of the upper bound theorem.

Assuming an associated flow rule, the angle of each cone is equal to φ. Thus, during collapse, the two adjacent edges of the mechanism remain in contact. Note that the intersection of a cone with the tunnel face is an ellipse (cf. Fig. 2.6)

Léca and *Dormieux* calculated the support pressure as a function of the inclination angle α of the cone axis with respect to the horizontal. By maximising the resulting pressure over a range of α, expressions for the necessary support pressure were obtained.

Given an **associated flow rule** (cf. Chap. 4), the direction of the plastic strain increment vector is normal to the yield surface. Thus, e.g. for a Mohr-Coulomb failure condition, the dilation angle ψ is equal to the friction angle φ.
In general, frictional soils do not obey an associated flow rule because $\psi < \varphi$. It was shown, though, that an upper bound calculated with $\psi = \varphi$ is also an upper bound for the case $\psi < \varphi$ (cf. [27]). *De Jong* [31] and *Palmer* [99] have shown that, under favourable conditions, lower bound solutions also hold for material with a non-associated flow rule.

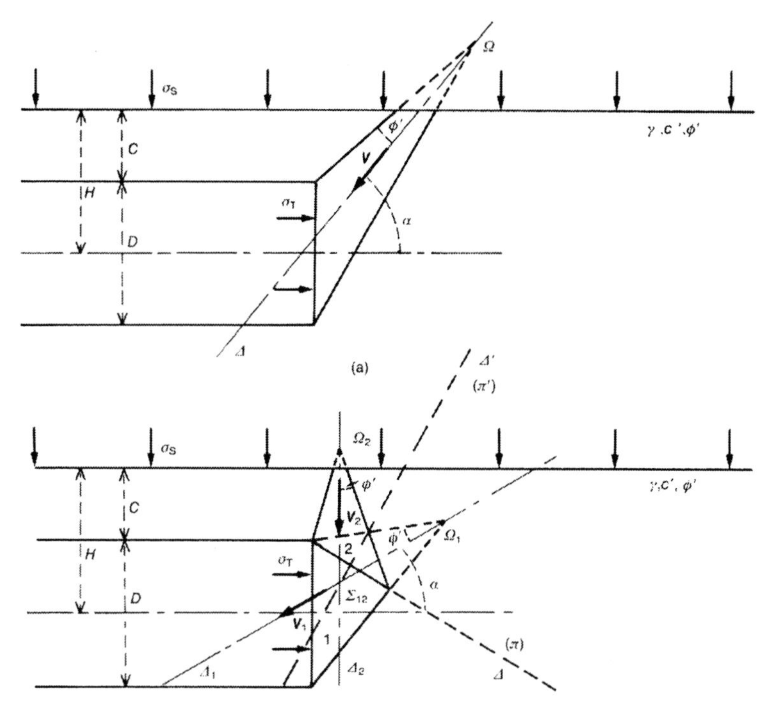

Figure 2.5: Failure mechanism with one (top) and two (bottom) conical wedges as assumed by *Léca* and *Dormieux*, [77]

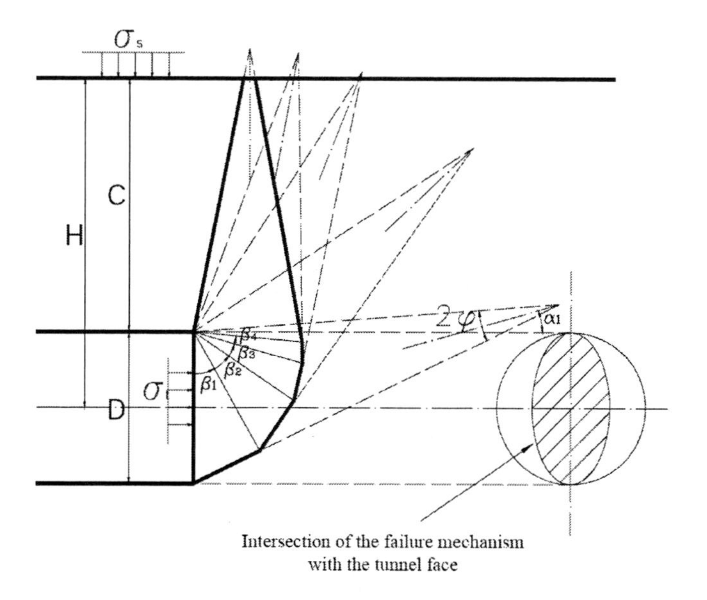

Figure 2.6: Failure mechanism with various conical wedges, proposed by *Soubra* et al. [114]

Soubra [112–114] improved the mechanism suggested by *Léca* and *Dormieux* by introducing a shear *zone* between the two conical wedges. This shear zone consists of n intermediate slices of conical wedges, whose surfaces form a log-spiral curve in a vertical cut of the problem (Fig. 2.6). Note that in contrast to the mechanism by *Léca* and *Dormieux*, *Soubra*'s rigid block mechanism has $n + 1$ free parameters: angles α and β_i $(i = 1 \ldots n)$.

Soubra stated that his mechanism leads to higher upper bound solutions for the given problem than the ones by *Léca* and *Dormieux*. In the framework of limit analysis, this is an improvement towards the true collapse load.

Moreover, *Soubra* found out that a total number of n = 4 slices is sufficient for the determination of the support pressure. A larger number of slices led to improvements smaller than 1 % of the final pressure [114].

2.1.2 Static approaches

For calculation of a **lower bound solution** within plasticity theory, a statically *admissible stress field* is chosen that fufils equilibrium and does not violate the yield condition in any point of the soil body. For the face stability problem a necessary support force is obtained that is higher or equal to the true collapse load F_{collapse}.
In this case, the accuracy of the solution depends on the similarity between the real and chosen stress fields. For further reference, see e.g. *Potts* and *Zdravković* [102] or *Mang* and *Hofstetter* [84].

Several authors (*Atkinson* and *Potts* [7], *Davis* et al. [29], *Mühlhaus* [91] or *Léca* and *Dormieux* [77]) have presented lower bound calculations for tunnelling problems in cohesive and non-cohesive soils.

Representative stress fields by ***Léca*** and ***Dormieux*** [77] are shown in Fig. 2.7. The two-dimensional stress field in Fig. 2.7 a represents a general case of a soil with weight ($\gamma > 0$). Although the three-dimensional stress field, shown in Fig. 2.7 b does not take the self-weight into account, it helps to improve the solution.

Summarising, the static approaches either neglect the self-weight of the soil or only treat the face-stability problem two-dimensionally in order to obtain a solution. Nonetheless, the lower bound solutions by *Léca* and *Dormieux* constitute a valuable reference for other approaches.

Smoltzcyk [111] also constructed stress fields around a circular tunnel in soil with friction, cohesion and self-weight. He expressed the three-dimensional stress state in the soil by means of a stress function of four independent material and geometric parameters. With particular assumptions for the stress field beyond the tunnel face

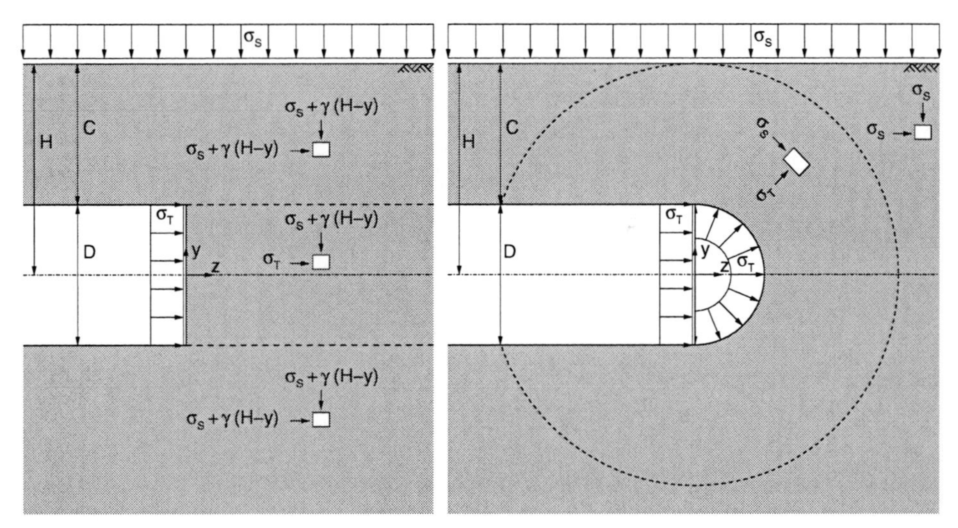

(a) Stress field for soil with self-weight (b) Stress field for soil without self-weight

Figure 2.7: Stress fields assumed by *Léca* and *Dormieux*, [77]

and application of a variational principle, *Smoltzcyk* obtained an expression for the horizontal stresses on the tunnel face.

For a shield heading in Hamburg, e.g., with $2 < C/D + 0.5 < 4$ and $\varphi = 30°$, *Smoltzcyk* put forward the following expression for the necessary support pressure p_f:

$$\frac{p_f}{\gamma\, D} = 0.237 + 0.17\, C/D \quad . \tag{2.1}$$

As this expression is only valid for $C/D > 1.5$, it was not considered further in this work.

2.1.3 Other theoretical approaches

A different approach, which is neither a failure mechanism nor a static solution, was presented by ***Kolymbas*** [73]. He assumed a parabolic distribution of vertical stress $\sigma_v(z) = az^2 + bz + c$ between the ground surface and the crown of the tunnel (Fig. 2.8). The tunnel face was assumed hemispheric.

Before excavation of the tunnel, σ_v varies linearly with depth: $\sigma_v(z) = \gamma z$.

With the assumptions:

- surcharge at surface level: $\sigma_v(z = 0) = q$,
- geostatic gradient of vertical stress at the surface level: $\sigma_v'(z = 0) = \gamma$,

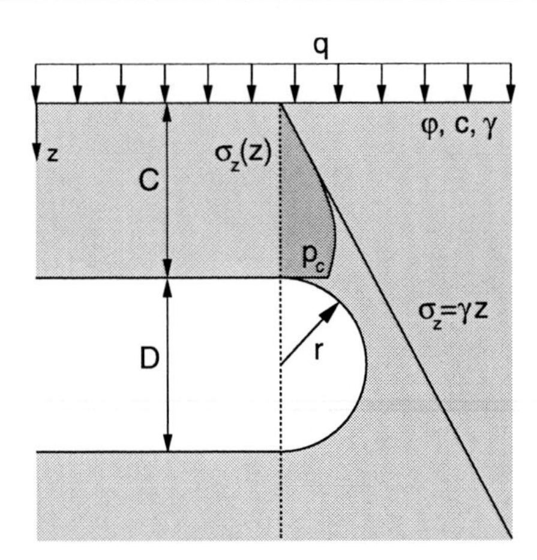

Figure 2.8: Vertical stress distribution assumed by *Kolymbas*, [73]

- full mobilisation of soil strength at the tunnel crown,

Kolymbas considered equilibrium at the tunnel crown. Thus, he obtained an equation for the necessary support pressure p_c at the crown, which serves as an estimate for the necessary support pressure p_f.

In a similar way, but with a hyperbolic approach for the vertical stress distribution *below* the tunnel, *Kolymbas* calculated the necessary support pressure p_i at the invert of the tunnel.

Balthaus [12] put forward two simple calculation methods, intended for engineering practice. The first formula approximates the necessary support pressure p_f^c at the crown of the tunnel as sum of water pressure p_w^c at the tunnel crown and averaged active horizontal earth pressure $\bar{e}_{ah}$ between ground surface and tunnel invert (Fig. 2.9).

The second scheme takes the horizontal earth pressure at the tunnel crown into account: p_f^c is approximated as sum of p_w^c and $k(\sigma_v' + p)$. In this expression k is an empirical earth pressure coefficient[2], σ_v' the effective vertical pressure at the tunnel crown and p a surcharge at ground level.

The disadvantage of both suggested formulas is the lack of mechanical background. *Balthaus* states that both approaches revealed a good agreement with the *Horn*-mechanism for an executed underground tunnelling project. But this statement does not hold in general.

[2] *Balthaus* suggests $k = 0.4$ for shallow tunnels.

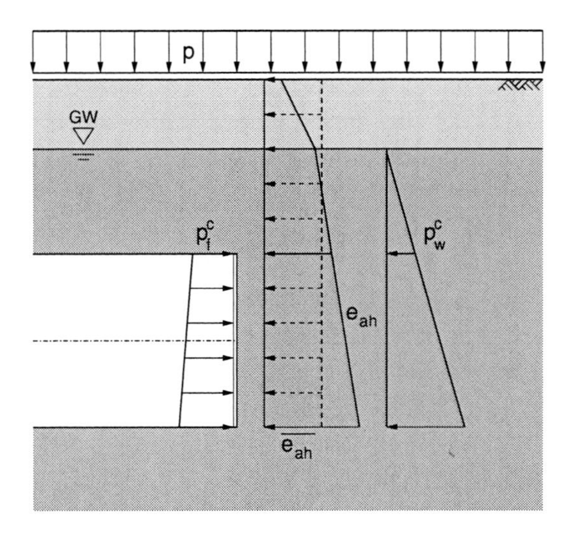

Figure 2.9: Illustration of the approximate formula by *Balthaus*, inspired by [12]

2.2 Experimental models

Researchers have been trying to model face stability problems experimentally, mainly with small-scale models. Full-scale experiments are very expensive and unsuitable for parameter studies, because their setup is not reproducible enough. Apart from that, safety concerns prevent access to tunnels which are close to collapse (cf. *Meguid* et al. [87]).

A distinction can be made between tests at single gravity, called *1g model tests*, and *centrifuge tests*:

- Tests under 1g conditions are less expensive than centrifuge tests and allow to study even complex problems under controlled conditions. They are easily reproducible so that a large number of tests can be performed, and parameter studies can be conducted. They have been used for a long time in experimental research.
  The major drawback of 1g tests is the low stress level which is not characteristic for real tunnelling problems. Moreover, it is difficult to determine the material properties at such low stresses.

- Some of the drawbacks of 1g tests are overcome with centrifuge modelling: due to the radial acceleration of the basket during flight, a small-scale model can be subjected to an elevated gravitation, ng (n denotes the multiple of the earth's gravity g). In doing so, the self-weight of the soil is scaled up, and the stress distribution with depth in the model becomes comparable to the stress distribution in a prototype, which is n times larger. The centrifuge experi-

ment allows, thus, to consider stress-dependent soil properties correctly. Centrifuge tests are favoured because they satisfy similarity conditions (*Chambon* and *Corté* [24]). The centrifuge technique has been applied since the 1930s to study geotechnical problems at universities, but it is increasingly applied in practice to support engineering design (c.f. e.g. *Laudahn* [76] or *Muir Wood* [92]).
However, there are limitations (cf. e.g. *Laudahn* [76]), e.g. centrifugal acceleration depends on the distance of a soil particle to the centre of the centrifuge; in other words, mass forces are not the same over the height of the model. The instrumentation of the model must be adapted to the model dimensions and is usually very sophisticated, as a data transfer during flight of the centrifuge is required. This is one reason why centrifuge tests are very expensive and require a long preparation time.

Meguid et al. [87] delivered a current state-of-the-art report on physical modelling of tunnels in soft ground. Also *Broere* [19] gave a detailed overview of laboratory and field observations on the matter. As this work is concerned with the face stability of shallow tunnels, only those studies that investigate the given problem will be presented here.

2.2.1 Centrifuge tests

Chambon and ***Corté*** [24, 25] conducted a series of tests with dry fine sand in a centrifuge. Fig. 2.10 shows their model of a fully lined tunnel. It consisted of a 10 cm diameter steel tube whose front was covered with a thin latex membrane that prevented soil from entering the cavity. A displacement transducer logged the horizontal displacement of the face, a pressure transducer measured the current pressure in the tunnel chamber. This pressure was either applied via air or water, taking thus constant and hydrostatic pressure distributions over the height of the face into account.

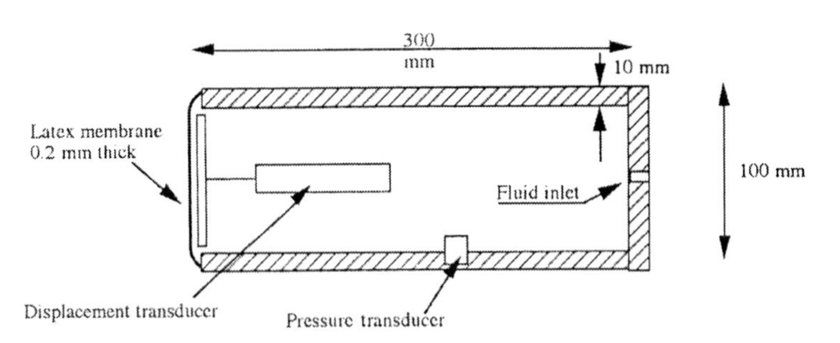

Figure 2.10: Schematic sketch of the tunnel cross section, [24]

An overview of the experimental setup is given in Fig. 2.11. The sand was pluviated into the box to achieve the best possible homogeneity. Relative densities between 0.65 and 0.92 were achieved in various tests. Via a water tank on top of the box, additional surcharges could be applied.

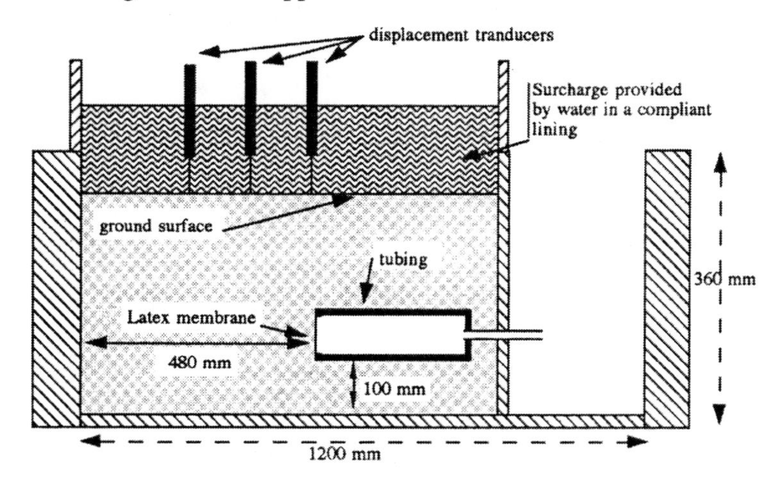

Figure 2.11: Experimental setup for centrifuge tests, [24]

Chambon and *Corté* varied the cover-to-diameter ratio C/D between 0.5 and 4.0, and they chose three acceleration levels for the centrifuge: 50g, 100g and 130g. Thus, they modelled prototype diameters of 5.0 m, 10.0 m and 13.0 m.[3] At the start of each test, the pressure inside the tunnel chamber was in equilibrium with the earth pressure acting on the latex membrane. Then the pressure inside the tunnel was gradually reduced until failure occurred. In this case, failure was defined as large increase of horizontal displacements without change in internal pressure. In addition, the shape of the failure mechanism was deduced from displacement patterns of coloured sand layers immediately after each test. For this purpose the sand was wetted and the soil body cut along different vertical planes.

A typical test result is plotted in a diagram of internal pressure vs. face displacement (Fig. 2.12). *Chambon* and *Corté* distinguish three stages of face collapse:

- *Phase 1* is characterised by a reduction of internal pressure without any face displacement. The stress at which displacements begin is found to be much lower than the active Rankine earth pressure (calculated by the authors for plane strain conditions).
- *Phase 2* marks a continuous increase in face displacement as reaction to a further pressure reduction in the tunnel. The authors point out that in this stage

[3]Lengths are scaled linearly with the acceleration level n; that means that a 10 cm diameter tunnel under 100g gravitation represents a prototype with 10 cm $\times 100$ = 10 m diameter.

holding or increasing the internal pressure brings the horizontal displacements to a stop.

- *Phase 3*, finally, characterises the collapse of the tunnel: at a pressure p_f the displacements increase rapidly. Therefore, this level of the internal pressure can be interpreted as necessary support pressure. The displacement at the onset of collapse was not much affected by the geometry of the model or the density of the sand.

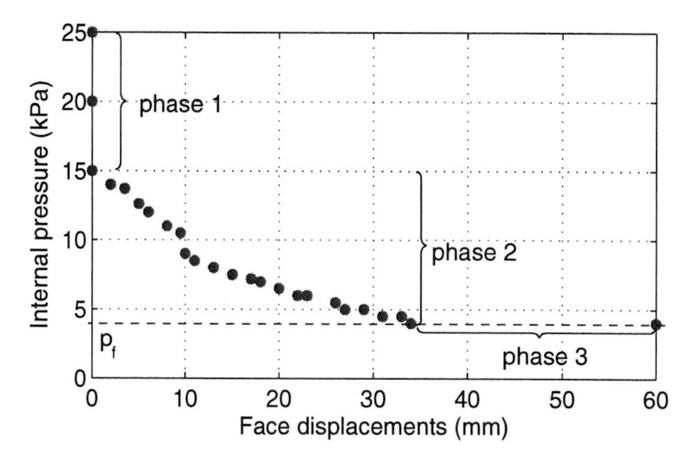

Figure 2.12: Increase of face displacement with decreasing confining pressure for C/D = 2.0, data from [24]

Chambon and *Corté* did not discover any relation between C/D and dimensionless support pressure $N_D := p_f/(\gamma\ D)$. There was no significant influence of initial density on N_D, either. The authors only mentioned that loose soils seemed to be less stable than dense soils. Also the type of support medium (air or water) did not affect N_D.

Chambon and *Corté* also presented sketches of the observed failure bulbs (Fig. 2.13), which were recorded *after* each test. These failure regions consist of a vertical chimney above the tunnel crown and are confined by a curved envelope in front of the tunnel face (*failure bulb*). Depending on the overburden C, the bulb closes at a certain depth due to arching above (for $C/D \geq 1.0$) or extends to the ground surface because a full arch could not develop (for $C/D \leq 0.5$).

The shape of the failure zone could only be observed at the end of each test. So, it is not possible to make a statement about the onset and propagation of collapse. The authors mentioned an influence of the type of support on the shape of the failure zone: whereas the whole face area was involved in the collapse for a pneumatic support, the observed displacements after tests with water as a support medium concentrated in the upper part of the face. *Chambon* and *Corté* point to the fact that in the latter case failure occurred suddenly when the water level inside the tunnel went below the tunnel-crown level.

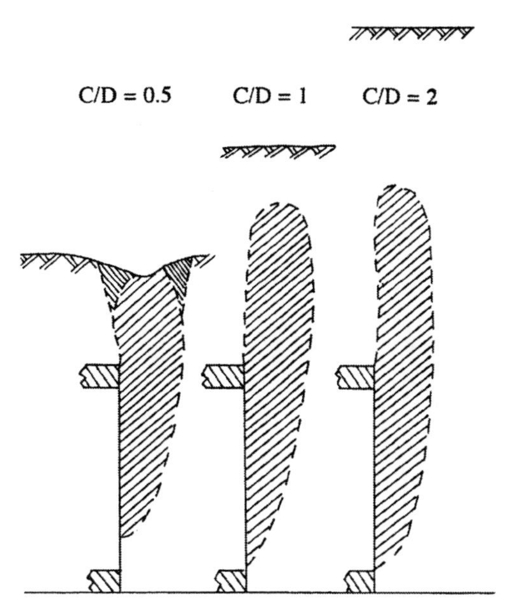

Figure 2.13: Shape of failure bulbs at different overburden ratios after each test (cut through axis of symmetry), [24]

Al Hallak et al. [2] obtained similar failure bulbs in centrifuge tests with an acceleration of 50g. The authors used dry Fontainebleau sand and $C/D = 2$. The failure zone at the end of their test is shown in Fig. 2.14.

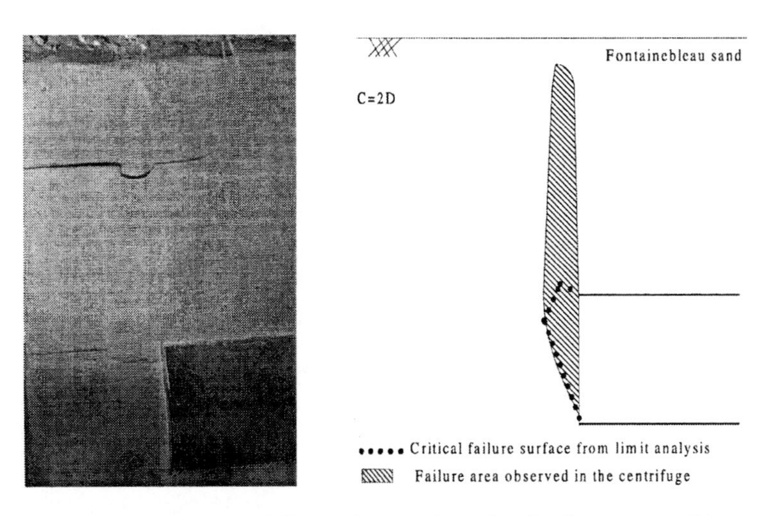

Figure 2.14: Shape of failure bulbs (cut through axis of symmetry), [2]

Plekkenpol et al. [101] investigated the face stability of a model of the Second Heinenoord Tunnel (The Netherlands) in the centrifuge. The geology of the prototype was modelled with fine sand (d_{50} = 0.16 mm) in three layers (Fig. 2.15 a). The sand was installed with relative densities between 0.47 (top layer) and 0.99 (bottom layer). In accordance with prototype conditions, the sand was fully saturated throughout the tests.

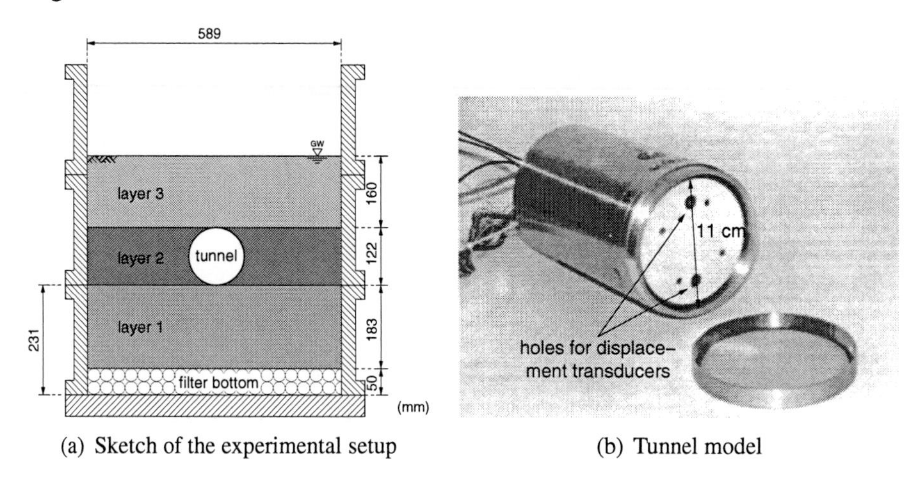

(a) Sketch of the experimental setup (b) Tunnel model

Figure 2.15: Model for centrifuge tests by *Plekkenpol* et al., [101]

The tunnel model consisted of a circular tube (inner diameter 11 cm) with a pressure chamber at the front (Fig. 2.15 b). This chamber was filled with support fluid and sealed with a flexible impermeable membrane at the face. The support pressure was controlled by pumping fluid into and out of the working chamber. Horizontal movements of the face were logged via displacement transducers at two locations (cf. Fig. 2.15 b).

The model was set up in steps: after preparation of the first soil layer the tunnel model was installed.[4] Then layers two and three were added and the centrifuge accelerated to the desired rotational speed[5]. Simultaneously, the pressure in the working chamber was increased to prevent movements of the tunnel face.

Failure of the tunnel face was triggered by pumping the support fluid out of the working chamber. Thus, the test was neither load- nor displacement-controlled; rather the soil volume entering the tunnel was controlled.

The resulting load-displacement curve (Fig. 2.16) shows similar characteristics to the one by *Chambon* and *Corté* (Fig. 2.12): after a sharp drop, the pressure reaches a minimum before it rises again with increasing displacement.

[4]During model setup the stability of the tunnel face was ensured by a mechanical support.

[5]*Plekkenpol* et al. did not mention the exact g-level of their tests. The author back-calculated a value of $n \approx 75$.

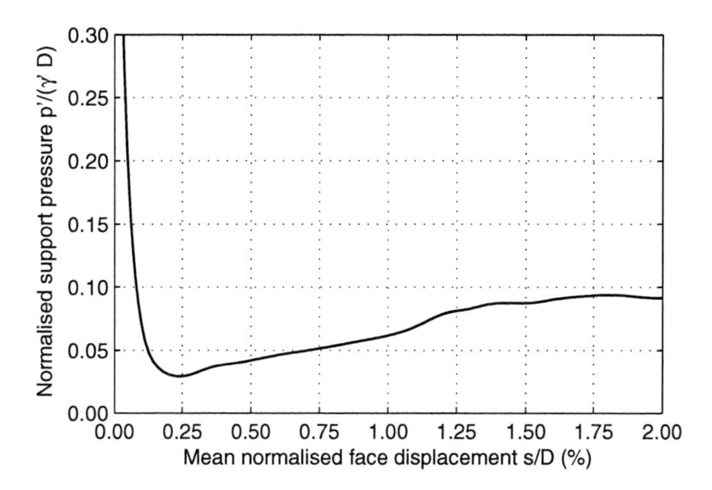

Figure 2.16: Load-displacement curve for centrifuge tests by *Plekkenpol* et al., data for p from [101], normalised with back-calculated values for $\gamma' \approx 9.8$ kN/m^3 and tunnel diameter D = 8.3 m

The face displacements were logged with two transducers installed 32.5 mm above and below the tunnel axis. The readings indicated that the displacements *above* the tunnel axis were larger than those *below* for the same stage of the experiment.

Plekkenpol et al. also made use of coloured sand layers to determine the extent of the failure zone after the test. Their experiments showed failure zones similar to those obtained by *Al Hallak* et al. or *Chambon* and *Corté*.

Kamata and ***Mashimo*** [62] also performed centrifuge tests with sand, using accelerations of 25g and 30g. They mainly investigated the influence of face bolts on face stability, but also presented results for the stability of an unbolted excavation face.

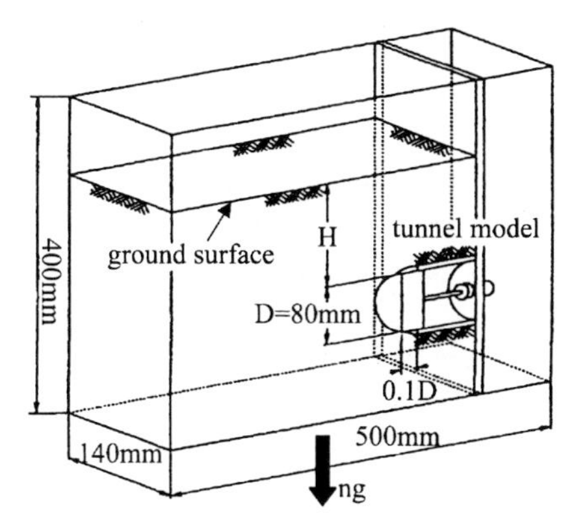

Figure 2.17: Experimental setup for centrifuge tests of *Kamata* and *Mashimo*, [62]

The problem was modelled in half, with an acrylic front panel to observe the soil movements (Fig. 2.17). The tunnel was represented by a semi-cylindrical shell with an inner diameter of 8.0 cm. The soil was supported by a movable semicircular aluminium plate that extended $P = 0.1\ D = 0.8$ cm into the soil domain. Thus, an unlined length in front of the tunnel face was modelled. Failure was induced by retracting the aluminium plate into the tunnel once the desired centrifugal acceleration was reached.

Kamata and *Mashimo* used unsaturated Toyoura sand (water content $w = 6.5\%$) with a cohesion of $c = 4.6$ kPa and a friction angle of $\varphi = 34.5°$.

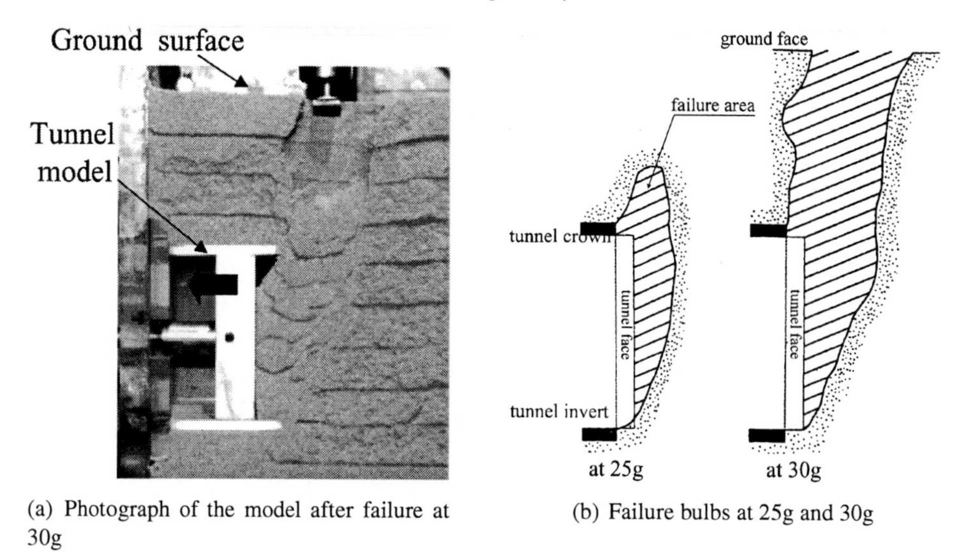

(a) Photograph of the model after failure at 30g

(b) Failure bulbs at 25g and 30g

Figure 2.18: Experimental results by *Kamata* and *Mashimo*, [62]

Fig. 2.18 shows the observed failure mechanisms: The failure zone for a centrifuge acceleration of 25g was termed *dome-shaped* [62]. It did not reach the ground surface. This happened, however, for an acceleration of 30g. The authors mentioned that in both cases a slip surface originated at the bottom of the tunnel face and spread upwards ahead of the face. The sketches indicate that there was no distinct failure mechanism detectable. *Kamata* and *Mashimo* made no further comment on the shape of the slip zone or the evolution of collapse.

Kimura and ***Mair*** [67] presented results of tests in soft clay that served to assess the necessary support pressure under undrained conditions. They developed a three-dimensional model which was cut vertically through the tunnel axis (Fig. 2.19). The front of the model box was transparent so that the behaviour of the clay could be observed throughout the test. The tunnel had an inner diameter of 6 cm. In a series of tests, the cover-to-diameter ratio C/D was varied between 1.5 and 3.0. Moreover, the unlined length P ahead of the tunnel was varied. The centrifuge generated accelerations of 75g and 125g, thus simulating prototype tunnel diameters of 4.5 m and 7.5 m.

During acceleration of the centrifuge, stability inside the tunnel chamber was supplied by a flexible rubber bag under compressed air. Then the supporting air pressure was reduced within a few minutes. When the necessary support pressure p_f was reached, displacements at the tunnel face increased rapidly.

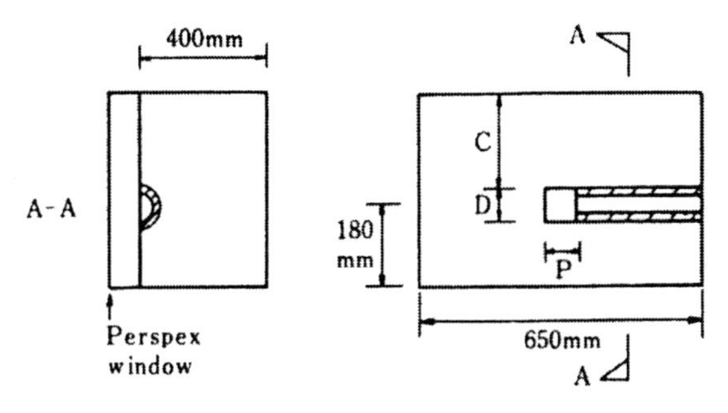

(a) Sketch of experimental setup

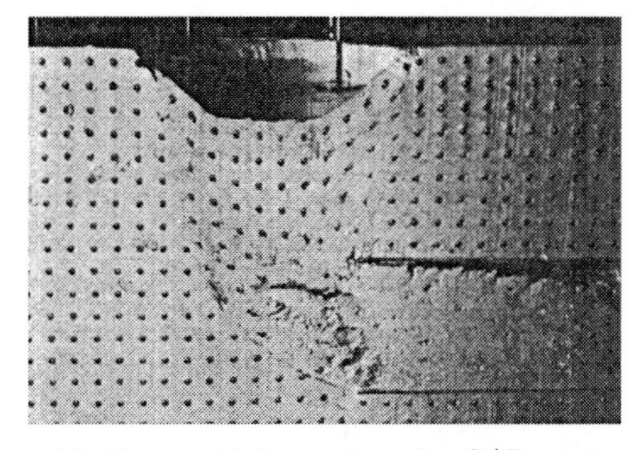

(b) Observed failure pattern for $C/D = 1.5$

Figure 2.19: Centrifuge model by *Kimura* and *Mair*, [67]

The observed tracer particle movements, such as those illustrated in Fig. 2.19 b, served to identify failure mechanisms. For undrained conditions, the necessary support pressure can be expressed in terms of the stability ratio N, proposed by *Broms* and *Bennermark* [20]:

$$N = \frac{\sigma_z - p_f}{c_u} \quad , \tag{2.2}$$

with the overburden pressure at the tunnel axis σ_z and the undrained shear strength c_u of the clay. Fig. 2.20 shows the relationship between N and C/D, obtained by *Kimura* and *Mair*. It indicates no increase of N for $C/D > 3$.

2.2.2 1g model tests

Takano et al. [121] investigated the effect of face bolting on the stability of a tunnel face in dry Toyura sand with a relative density of 0.8 under 1g conditions. Their experimental setup is shown in Fig. 2.21: the tunnel was modelled with a hollow

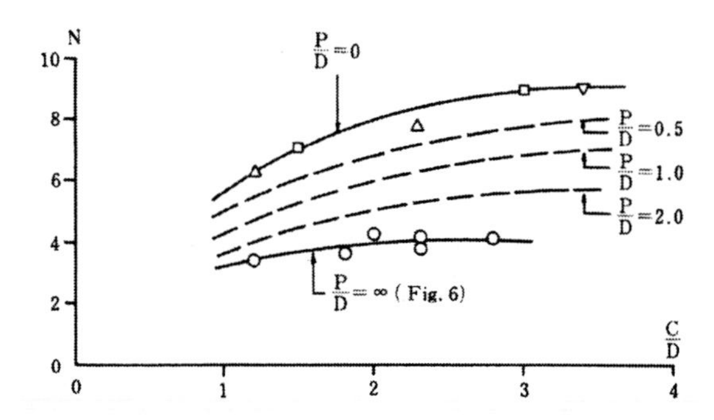

Figure 2.20: Relation between stability ratio at failure N and cover-to-diameter ratio C/D, [67]

cylinder with a diameter of 2 cm. The overburden was 4 cm, i.e. $C/D = 2$. The tunnel face was supported by a solid core which was pulled out via an electric motor during the test. After a piston displacement of 2 mm the test was stopped, and the whole model was moved to an X-ray CT.

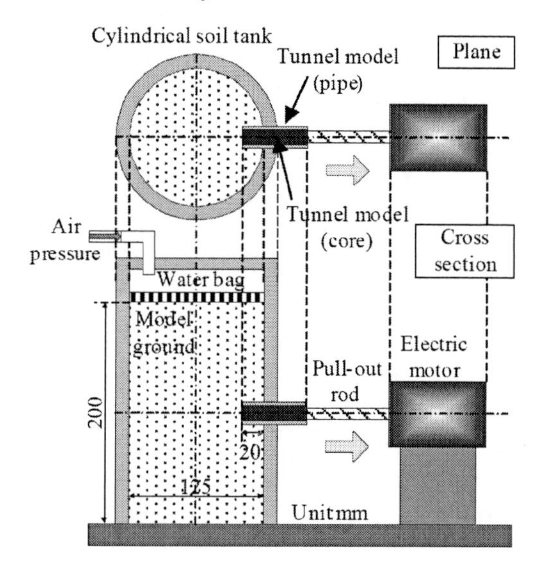

Figure 2.21: Experimental setup of *Takano* et al., [121]

The soil body was scanned in layers of 1 mm thickness from bottom to top, and three-dimensional wireframe plots were generated from the obtained density distributions. The failure body is visible in Fig. 2.22 a as light grey zone of increased porosity. The wireframe plot (Fig. 2.22 b) shows the three-dimensional structure of the mechanism, illustrating the border between dense and loosened soil. The shape of the zone of dilation is in good correlation with experimental results presented in Sec. 2.2.1.

Sterpi and ***Cividini*** [116] experimentally and numerically investigated the behaviour of shallow tunnels in dense sand under 1 g conditions. Their three-dimensional model (Fig. 2.23 a) included a 1.1 m long horse shoe shaped tunnel, 1.32 m wide and

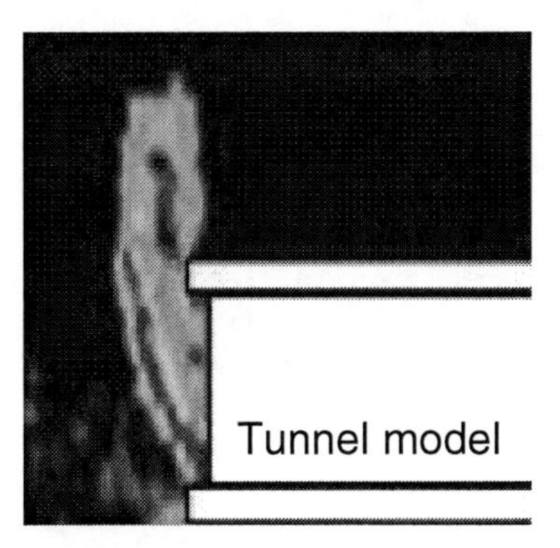

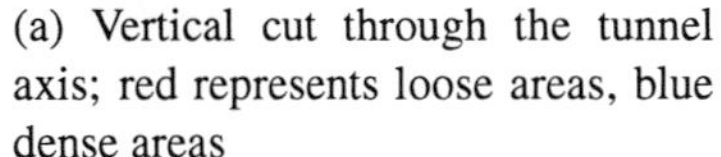
(a) Vertical cut through the tunnel axis; red represents loose areas, blue dense areas

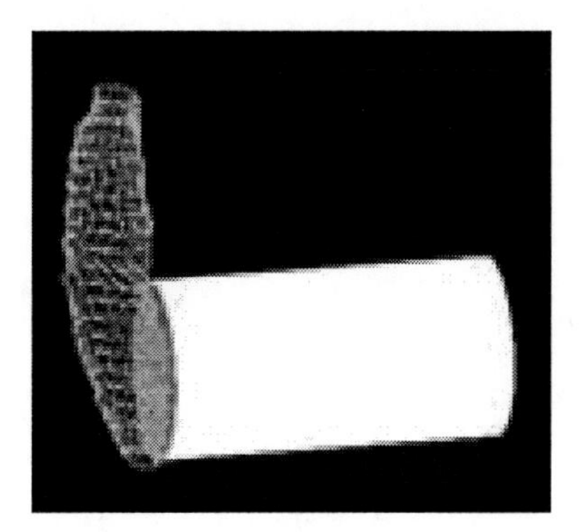
(b) Wireframe plot

Figure 2.22: Failure zone as detected by X-ray CT, [121]

1.145 m high. The overburden at the tunnel crown was equal to the tunnel width. Support was supplied via a vinyl bag filled with pressurised air that was placed inside the excavation chamber. Failure was triggered by lowering the air pressure inside the chamber.

Sterpi and *Cividini* used dry medium uniform sand with a peak friction angle $\varphi_p = 32°$ and a residual friction angle $\varphi_r = 32°$ which was pluviated into the model container. Displacements inside the soil were logged with transducers and inclinometers, and evaluated in given time intervals.

From the obtained measurements *Sterpi* and *Cividini* concluded that a shear band developed at the centre of the tunnel invert which then spread towards the sides and ahead of the excavation face. Finally, it headed towards the surface in a nearly vertical direction. A complementary zone of shear localisation started from the tunnel crown and extended vertically to the ground surface. Collapse finally occurred when both shear zones isolated the soil volume depicted in Fig. 2.23 b. At this stage the pressure at the tunnel face was about 15 to 20% of its initial value.

2.3 Numerical models

During the last 20 years or so, the number of numerical studies on the ground response to tunnelling has increased significantly. Different numerical approaches have been applied; both theoretical and practical aspects of the given problem have been tackled.

In the following overview the focus lies on the face stability of shallow tunnels in soft ground, preferably with investigations of necessary support pressure or failure

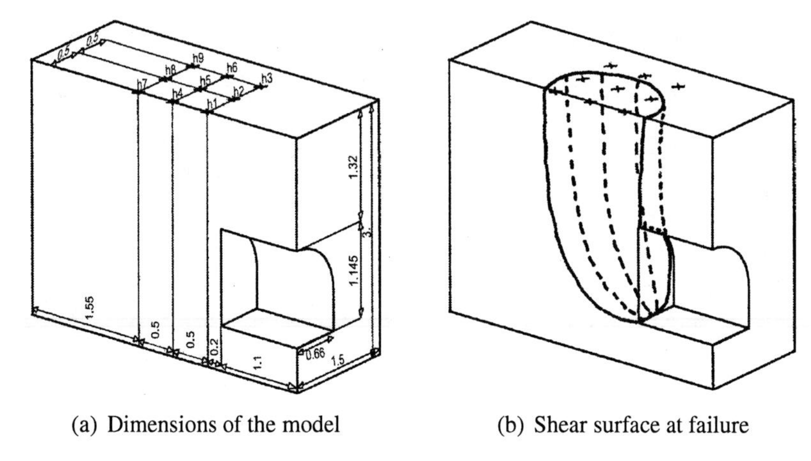

(a) Dimensions of the model (b) Shear surface at failure

Figure 2.23: Experimental setup and results by *Sterpi* and *Cividini*, [116]

patterns. Other aspects of tunnelling in soft ground, such as modelling of the construction process, interaction between soil and lining or time-dependent material properties will not be covered here. An extensive overview of the application of numerical methods to tunnelling problems was given by *Gioda* and *Swoboda* [44].

2.3.1 Finite element analysis

In recent years the **F**inite **E**lement **M**ethod (FEM) has become an important tool for the simulation of engineering structures, because it allows the modelling of complex structures with advanced constitutive models. Complex ground conditions, irregular cross sections or the soil-structure interaction can be taken into account. The FEM has been used for a wide range of tunnelling problems – those studies that focus on face stability will be presented here. A good overview of three-dimensional FE-calculations for tunnelling problems in general is given by *Gioda* and *Swoboda* [44] or *Bliem* [16].

As the stability of the excavation face is a three-dimensional problem, only three-dimensional models are able to reproduce realistic stress and strain states in the ground. Yet, in engineering practice, mostly two-dimensional models are used, because they require less computation time and the modelling process is easier. In addition, practicable numerical tools for three-dimensional modelling have only been available for the engineering practice since the 1990s.

Ruse and ***Vermeer*** [107, 130, 132] investigated the necessary support pressure for the tunnel face of shallow tunnels with finite elements. They used 15-node prismatic volume elements for the soil and shell elements for the tunnel lining. An example for a finite element mesh they used is given in Fig. 2.24 a.

The soil was modelled with a linear elastic, perfectly plastic constitutive model with a Mohr-Coulomb failure condition. The effect of associated and non-associated flow was also investigated. *Ruse* pointed out that the Mohr-Coulomb model is well suited for the prediction of collapse loads, when pre-failure deformations are not important [130, 132]. Failure was triggered by a stepwise reduction of the support pressure p on the tunnel face.

The relation between p and displacement of a reference point (centre of the tunnel face) is illustrated in Fig. 2.24 b. Thus, *Ruse* reproduced typical curves that were observed in experimental campaigns (cf. the study by *Chambon* and *Corté*): with decreasing support pressure p the displacement of the face increased and headed towards infinity for $p = p_f$.

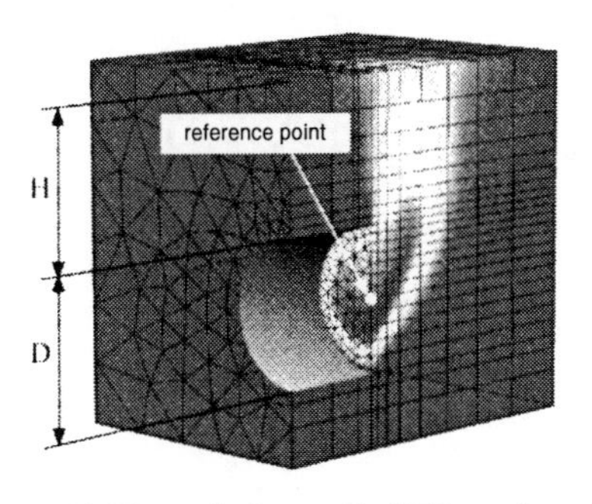

(a) Example for applied FE-mesh

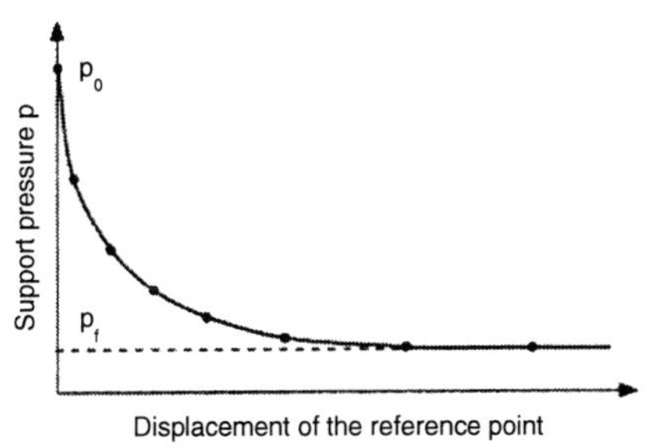

(b) Numerical load-displacement curve for the reference point

Figure 2.24: Numerical setup and results by *Vermeer* et al., [132]

Ruse investigated the influence of the following parameters on the necessary support pressure:

- initial ratio of lateral to vertical stress, $K := \sigma_h/\sigma_v$,
- elastic parameters: Young's modulus E and Poisson's ratio ν,
- material parameters for the plastic behaviour: friction angle φ, dilation angle ψ and effective cohesion c'.

He found out that K influences the magnitude of the displacements for a given p, but *not* the collapse load.[6] The same statement holds for the elastic parameters E and ν: those only influenced the displacements of the soil, but not p_f. Also the dilation angle had negligible influence on p_f; calculations with an associated flow rule ($\psi = \varphi$) and a non-associated flow ($\psi = 0$) led to the same p_f.

[6]Therefore, *Ruse* chose an earth pressure coefficient $K_0 = 1.0$ for all calculations.

Ruse expressed the necessary support pressure p_f in analogy to the bearing capacity equation as a function of dimensionless factors:

$$p_f = -c\, N_c + \gamma\, D\, N_D + q\, N_q$$

incorporating the cohesion c, the self-weight of the soil γ, the diameter of the tunnel D and a surface surcharge q. N_c, N_D and N_q are dimensionless coefficients.

He observed that the necessary support pressure only depended on the friction angle and the cohesion: for $\varphi \geq 20°$, arching developed within the soil body and made p_f independent of C/D (Fig. 2.25).

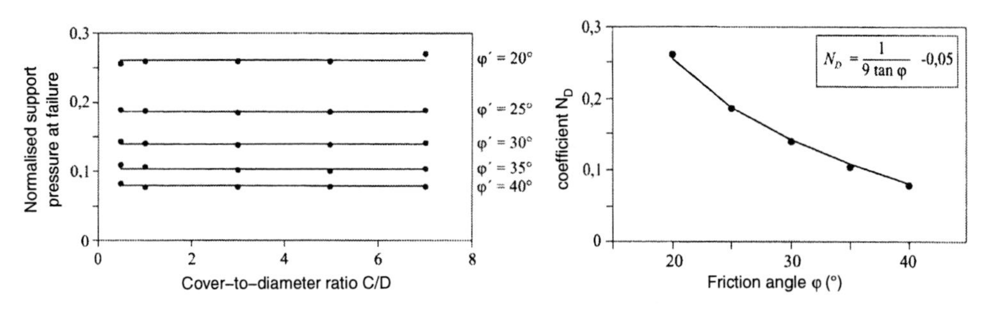

(a) Numerically predicted support pressure for a variation of C/D and φ ($c = 0$ and $\psi = 0$) (b) Assumed relationship between N_D and φ

Figure 2.25: Relation between normalised support pressure $p_f/(\gamma\, D)$, cover-to-diameter ratio C/D and friction angle φ, [107]

Ruse suggested an additional term for the calculation of p_f to take cohesion into account: $p_f = f(\varphi) - c\, \cot\varphi$. He found this approach confirmed by a series of finite element calculations (Fig. 2.26).

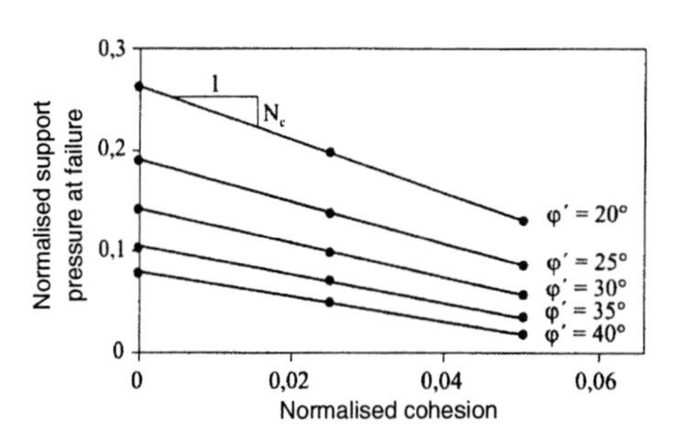

(a) Numerically predicted support pressure for a variation of φ and c

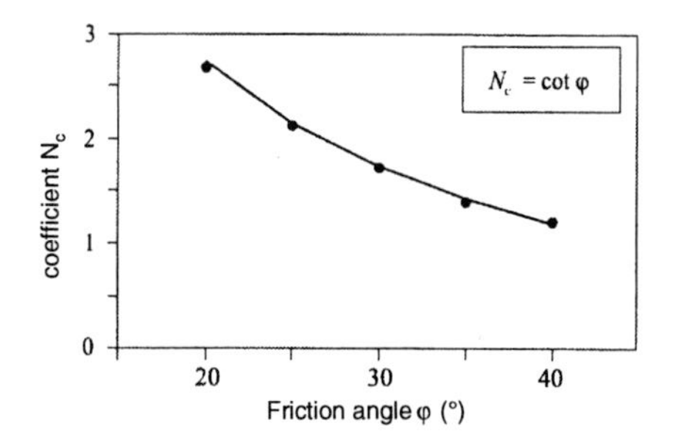

(b) Assumed relationship between N_c and φ

Figure 2.26: Relation between normalised support pressure $p_f/(\gamma\, D)$, normalised cohesion $c/(\gamma\, D)$ and friction angle φ, [107]

From his numerical parametric study *Ruse* suggested the following relation between p_f and the parameters φ and c:

$$p_f = \gamma\, D \left(\frac{1}{9\, \tan\varphi} - 0.05 \right) - c\, \cot\varphi$$

Besides the determination of p_f, the finite element simulations gave an insight into the displacement pattern of the soil body during decrease of the support pressure. *Vermeer* and *Ruse* [132] presented plots (Fig. 2.27) of the principal stresses and incremental displacements for calculations with different φ and $\psi = 0$.

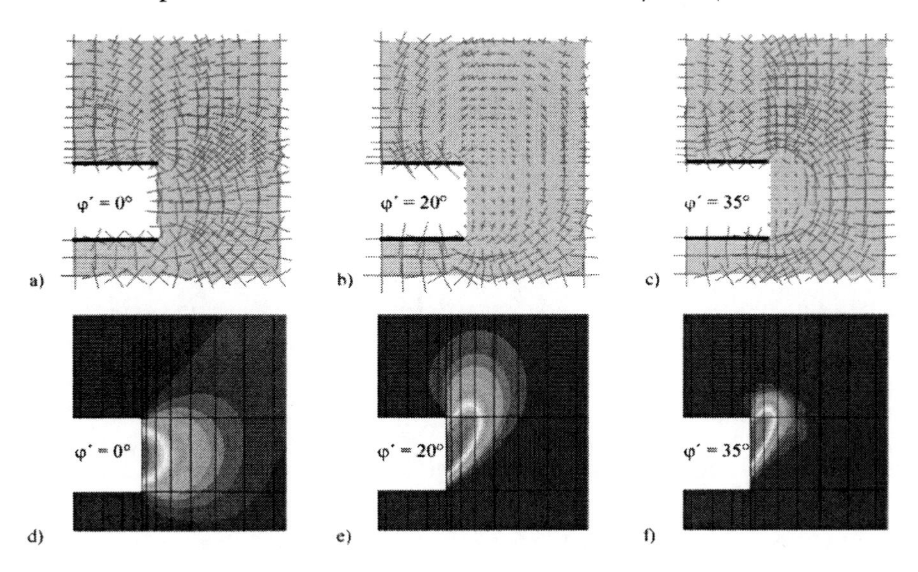

Figure 2.27: Principal stresses (a-c) and incremental displacements (d-f) close to failure. Close-up around the face for a tunnel with $C/D = 5.0$, [132]

The authors came to the conclusion that arching only developed for $\varphi \geq 20°$. The shape of the failure zone is in good correlation with experimentally observed ones (e.g. Fig. 2.18 on page 28 or Fig. 2.22 on page 31).

These results are supported by *Schubert* and *Schweiger* [109], who obtained similar results in comparable finite element calculations.

Sterpi and ***Cividini*** [116] conducted a finite element study for the face stability of a shallow tunnel as complement to their experimental investigation. They modelled the three-dimensional problem with 8-noded brick elements and a material model that incorporated a reduction of shear strength with increasing deviatoric plastic strains.

Due to symmetry, the problem domain could be modelled in half, cutting vertically through the tunnel axis. The steps of the simulation consisted of an initial geostatic equilibrium step with an earth pressure coefficient of $K = 0.5$, before the nodal forces on the excavation face were gradually reduced until failure occurred.

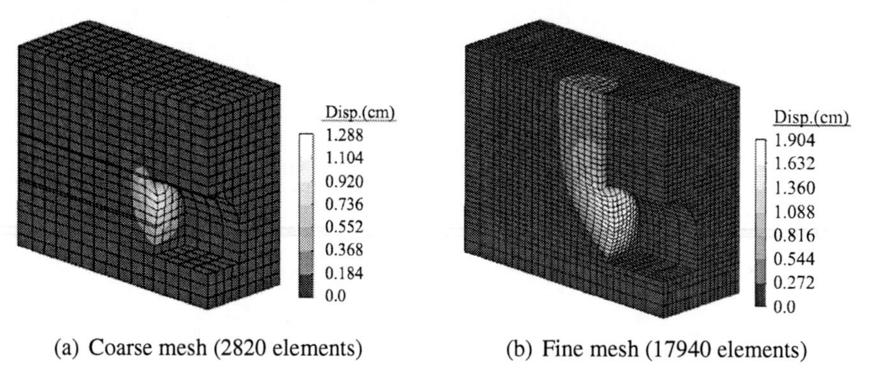

(a) Coarse mesh (2820 elements) (b) Fine mesh (17940 elements)

Figure 2.28: Displacement patterns for numerical calculations with consideration of strain softening, [116]

Sterpi and *Cividini* found a pronounced influence of element sizes on the obtained results. For a coarse mesh (2820 elements) the failure zone did not extend much into the soil, whereas for a finer mesh (17940 elements) it reached up to the ground surface (Fig. 2.28). The calculated displacements and the resulting support pressure for the coarse mesh did not reflect their experimental results. *Sterpi* and *Cividini* suggested that the "simple" brick elements have a limited capacity to capture the sharp variations in the strain field close to the tunnel face.

The authors made additional calculations with a linear elastic, perfectly plastic material model with Drucker-Prager yield condition, which could not consider strain softening effects. From the obtained results they concluded that a neglect of strain softening leads to an underestimation of displacements and, more important, necessary support pressure.

Mayer et al. [86] investigated the face stability with a three-dimensional finite element calculation. They applied a linear elastic, perfectly plastic material model with Mohr-Coulomb failure condition. A non-associated flow rule was applied ($\varphi = 20°$, $\psi = 8°$).

Failure was triggered by the *shear strength reduction technique*: $\tan\varphi$ and c were linearly and simultaneously reduced until failure occurred. The underlying safety factor is defined as the ratio of design values and material parameters at failure:

$$\eta = \frac{\tan\varphi_d}{\tan\varphi_f} = \frac{c_d}{c_f}$$

Mayer et al. modelled a non-circular tunnel with three-dimensional continuum elements, the shotcrete lining was also modelled with continuum elements (Fig. 2.29 a).

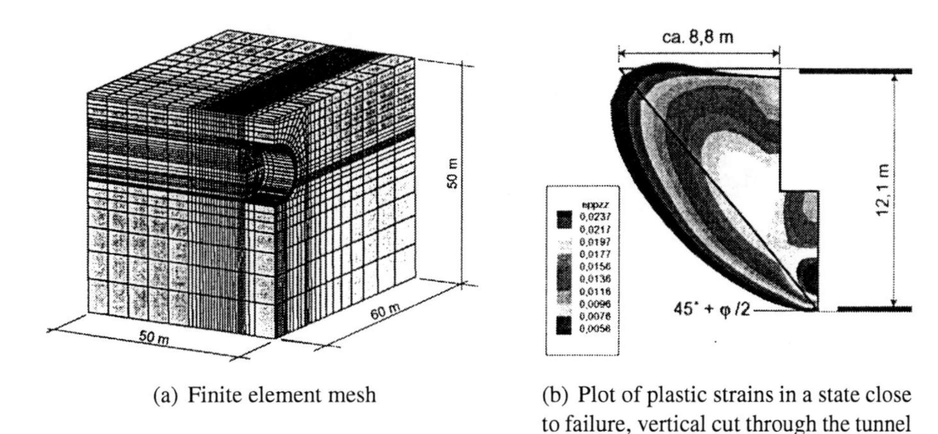

(a) Finite element mesh

(b) Plot of plastic strains in a state close to failure, vertical cut through the tunnel axis

Figure 2.29: Numerical discretisation and results by *Mayer* et al., [86]

Mayer et al. calculated displacements at failure, which were in good correlation with experimental results for the stability of diaphragm walls, e.g. by *Karstedt* [64]; they show the wedge-chimney structure, proposed by *Horn*. Still, the failure surface in the FE calculations was not a straight line, but curved (Fig. 2.29 b).

In a subsequent parametric study the overburden C and the cohesion of the soil c were varied: the authors concluded that the safety factor η did neither depend on C nor on c for $C/D \geq 1.5$. For simulations with a lower overburden a higher cohesion increased the safety factor.

Also the influence of the dilation angle ψ on η was studied, varying ψ between 0 and 8°. Whereas the obtained safety factor was hardly affected for a tunnel with $C/D = 5.7$, ψ had a significant influence on η for a shallow tunnel with $C/D = 0.5$: the resulting safety factors varied between $\eta_{\psi=0} = 1.69$ and $\eta_{\psi=8^\circ} = 2.47$.

Chaffois et al. [23] presented results of a three-dimensional FE calculation for shield tunnelling of a shallow tunnel in cohesionless gravel. They made use of an elasto-plastic material model with work-hardening characteristics. During the analysis the nodes forming the tunnel lining were fixed, and the pressure on the tunnel face gradually reduced.

The authors obtained a characteristic relation between support pressure and relative displacement of the face (Fig. 2.30). The linear part of the curve represents a purely linear elastic material response, the curved part indicates that stress points in the vicinity of the face reached the yield condition. In addition, the numerical simulations revealed arching in front of the face.

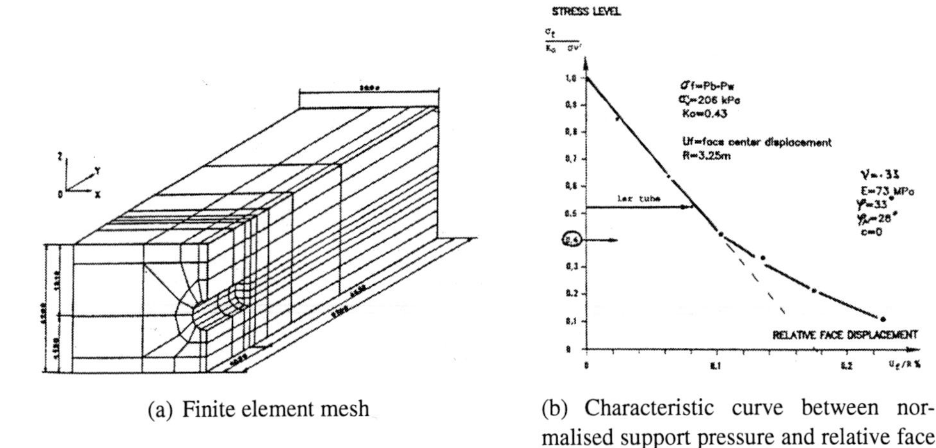

(a) Finite element mesh

(b) Characteristic curve between normalised support pressure and relative face displacement

Figure 2.30: Numerical discretisation and results by *Chaffois* et al., [23]

2.3.2 Discrete element analysis

The **D**iscrete **E**lement **M**ethod (DEM) allows to model the soil as a discontinuous medium. For this purpose, discrete blocks or particles are used, which interact on contact surfaces. Each block is treated as deformable material, but, in contrast to the FEM, separation of blocks is possible. The inherent disadvantage of standard DEM applications is that the choice of block geometry determines the position of possible slip surfaces.

The two-dimensional DEM was applied by ***Kamata*** and ***Mashimo*** [62] to analyse face stability.

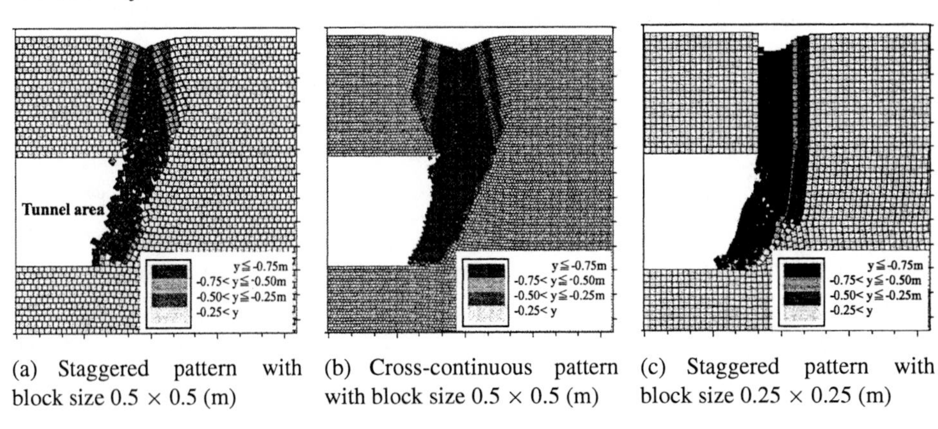

(a) Staggered pattern with block size 0.5 × 0.5 (m)

(b) Cross-continuous pattern with block size 0.5 × 0.5 (m)

(c) Staggered pattern with block size 0.25 × 0.25 (m)

Figure 2.31: Block displacement pattern for the UDEC-analysis, [62]

For their analysis, the block sizes and the block arrangement as well as the joint strength parameters (i.e. φ and c) were varied. Fig. 2.31 shows the obtained displacement patterns. The authors mentioned that during the analyses a slip surface evolved from the bottom of the face and extended towards the ground surface. *Kamata* and *Mashimo* defined failure as the onset of block separation.

Figs. 2.31 a and 2.31 b show clearly that the displacement pattern depended strongly on the initial arrangement of blocks. A reduction of block size (Fig. 2.31 c) did not change the final displacement pattern much, as *Kamata* and *Mashimo* mentioned. Moreover, computing power and time put a practical limit to the minimum block size. But *Kamata* and *Mashimo* pointed out that the advantage of discrete modelling over finite element analysis is the ability to judge better when collapse occurs.

The doubt remains that the three-dimensional face stability problem can be tackled satisfactory with a two-dimensional approach.

2.4 Case studies

Unfortunately, published data for face instabilities or even a collapse of a real-scale tunnel project are rare. To the author's knowledge, the only available data were published by *Jancsecz* and *Steiner* [60] for the Grauholz tunnel in Switzerland. During construction, the pressure in the excavation chamber was gradually reduced until instabilities occurred. The tunnel section in question (D = 11.6 m) was headed in fluvioglacial gravel with a friction angle $\varphi \approx 38°$ and a self-weight of γ = 22 kN/m^3. The authors mentioned support pressures at the onset of failure between 15 and 25 kPa, corresponding to N_D values between 0.06 and 0.10.

There are published data for *applied* support pressures of tunnel projects, such as the Grauholz tunnel close to Bern in Switzerland (*Jancsecz* and *Steiner* [60]), a tube tunnel in Düsseldorf in Germany (*Jancsecz* et al. [58]), the underground and surface railway system in Izmir in Turkey (*Jancsecz* et al. [59]), the Second Heinenoord tunnel close to Rotterdam in the Netherlands (*Bakker* et al. [9, 10]), the Botlek tunnel in the Netherlands (*Bakker* et al. [10]) and various projects in Japan (*Kanayasu* et al. [63]), to name just a few. But the calculation methods to arrive at the quoted support pressures are hardly ever mentioned; exceptions are listed in Sec. 2.6.

2.5 Comparison of existing models

The previous sections have shown that existing approaches are manifold – the motivation for this work was to deepen the understanding of underlying mechanisms and support or refute proposed ideas.

2.5.1 Example calculation

A comparison of some selected models was made by means of a simple example calculation (cf. also *Kirsch* and *Kolymbas* [68]). The dimensionless factor $N_D = p_f/(\gamma\ D)$ was calculated for the following set of geometry and material parameters:

- self-weight γ = 18 kN/m^3,
- cohesion $c = 0$,
- tunnel diameter D = 10 m,
- no surcharge on the ground surface,
- no groundwater.

The following models were compared:

- variations of the *Horn* model (theoretical, upper bound),
- *Krause* (theoretical, upper bound),
- *Léca/Dormieux* (theoretical, upper bound),
- *Kolymbas* (theoretical),
- *Ruse/Vermeer* (empirical, from FE calculations).

The *Horn* model was included because it is frequently used in engineering practice (cf. Sec. 2.6). The others represent different theoretical/empirical approaches.

In a first step, the cover-to-diameter ratio was C/D = 1.0, and the friction angle φ was varied between 25 and 45° (Fig. 2.32). In a second step, φ was kept constant at 32°, while C/D was changed between 0 and 3 (Fig. 2.33).

It was already mentioned that the *Horn* model allows for different combinations of lateral earth pressure coefficients, distribution of lateral earth pressure on the sides of the wedge, consideration or not of horizontal forces on top of the wedge and geometry of the wedge (Sec. 2.1.1). A total number of 108 *Horn* configurations is presented here.[7]

All models, reasonably, predict decreasing necessary support pressures for an increase of φ (Fig. 2.32). But for a given friction angle of, say, φ = 32°, the model responses vary between $N_D \approx 0$ and $N_D \approx 0.23$.

For a fixed φ and an increase in C/D the model responses (Fig. 2.33) are also manifold: *Kolymbas* and some *Horn* variations predict an increase of N_D, some *Horn* variations a decrease of N_D and some predictions are insensitive to a change of C/D.

[7] 108 configurations = K_{silo} (3 options) $\times$ K_{wedge} (2 options) $\times$ earth pressure distributions (3 options) $\times$ consideration of C_1 and $Q_{1,h}$ or not (2 options) $\times$ approximation of the tunnel area (3 options)

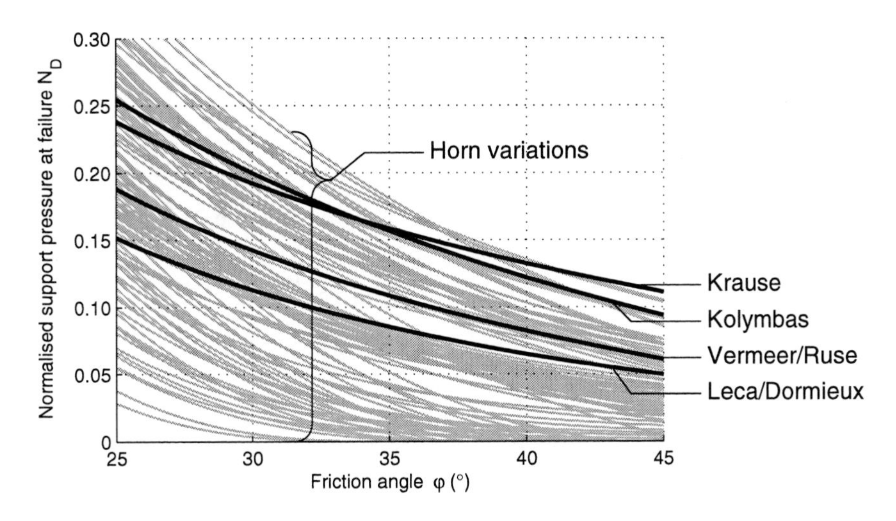

Figure 2.32: Relation between N_D and φ for different models, C/D = 1.0

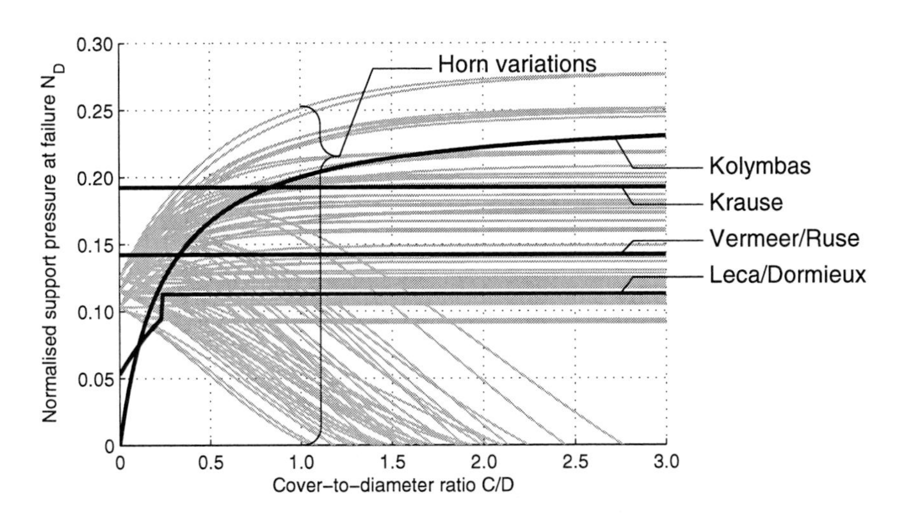

Figure 2.33: Relation between N_D and C/D for different models, φ = 32°

The decrease of some of the *Horn* predictions for an increase of C/D (Fig. 2.33) can be explained with the choice of earth pressure distribution on the sides of the wedge: if this distribution is linear from the ground surface downwards, the resisting frictional forces on the sides of the wedge become large and make additional support unnecessary for higher depths. This prediction is rather unrealistic, because in cohesionless material there will always be need for some support, even at larger C/D ratios.

Especially the comparably large variations within the *Horn* model suggest that its application is not straightforward, but results are rather sensitive to changes in the model configuration.

Summarising the comparison, the choice of *the* appropriate model for calculation of the necessary support pressure is not easy. Some models predict support pressures that might be too high. This could lead, in an extreme case, to blow-outs or heave at the ground surface (as reported by *Holzhäuser* et al. [54]). With other models the calculated support pressure might easily be unsafe, and the tunnel could collapse.

2.5.2 Sensitivity analysis

Apart from the deterministic result for a given set of input parameters, the uncertainty of the model responses can serve as quality criterion for the different approaches. The uncertainty of a variable x is usually expressed in terms of standard deviation σ_x or coefficient of variability

$$V_x = \frac{\sigma_x}{\mu_x}$$

with the expected value μ_x.

The input parameters for the proposed models are statistically distributed quantities with corresponding uncertainties. E.g. the friction angle φ of sand is assumed to have $V_\varphi \approx 0.1$, i.e. an uncertainty of roughly 10% (*Harr* [49]).

How do different models propagate the uncertainties of the input parameters? How sensitive are the approaches with respect to the input parameters? To answer these questions, a sensitivity analysis of the chosen models was performed.

Mathematical background

Consider a model for the determination of the necessary support pressure as a sufficiently smooth function f depending on random variables x_i $(i = 1, \ldots, n)$ with corresponding nominal (mean) values μ_i and standard deviations σ_i. By means of first order Taylor-series expansion, f can be linearised at a point $\boldsymbol{\mu} = \{\mu_1, \mu_2, \ldots, \mu_n\}$:

$$f \approx f(\boldsymbol{\mu}) + \sum_{i=1}^{n} \left. \frac{\partial f}{\partial x_i} \right|_{\boldsymbol{\mu}} (x_i - \mu_i).$$

The variance of f, σ_f^2, can then be expressed as

$$\sigma_f^2 = \sum_{i=1}^{n} \left(\left. \frac{\partial f}{\partial x_i} \right|_{\boldsymbol{\mu}} \right)^2 \sigma_i^2 + 2 \sum_{i=1}^{n} \sum_{j=1}^{i-1} \left. \frac{\partial f}{\partial x_i} \right|_{\boldsymbol{\mu}} \left. \frac{\partial f}{\partial x_j} \right|_{\boldsymbol{\mu}} \sigma_{ij} \quad , \tag{2.3}$$

parameter	**mean value** μ_i	**coefficient of variability** V_i	**standard deviation** σ_i	*source*
self unit weight γ	18 kN/m^3	2.0 %	0.36 kN/m^3	[49]
friction angle φ	32°	10.0 %	3.2°	[49]
tunnel diameter D	10 m	0.2 %	0.02 m	*assumption*
tunnel cover C	10 m	1.0 %	0.10 m	*assumption*

Table 2.1: Values for input parameters

with possible covariances σ_{ij} between the input variables.[8]

If the input parameters x_i are uncorrelated, (2.3) reduces to:

$$\sigma_f^2 = \sum_{i=1}^{n} \left(\left. \frac{\partial f}{\partial x_i} \right|_{\boldsymbol{\mu}} \right)^2 \sigma_i^2 \quad , \tag{2.4}$$

or in dimensionless form:

$$\begin{aligned} V_f^2 = \left(\frac{\sigma_f}{\mu_f} \right)^2 &= \sum_{i=1}^{n} \left(\left. \frac{\partial f}{\partial x_i} \right|_{\boldsymbol{\mu}} \frac{\mu_i}{\mu_f} \frac{\sigma_i}{\mu_i} \right)^2 \\ &= \sum_{i=1}^{n} \Big(\underbrace{\left. \frac{\partial f}{\partial x_i} \right|_{\boldsymbol{\mu}} \frac{\mu_i}{\mu_f}}_{=:\, k_i} V_i \Big)^2 \quad . \end{aligned} \tag{2.5}$$

The k_i are the dimensionless sensitivities of function f at point $\boldsymbol{\mu}$ with respect to the input parameters x_i. They allow the assessment of the "direction" of a cautious estimate for a parameter and the comparison of the impact of parameter variations on the result of f.

There is little information on the distribution, variances and covariances of the applied input parameters. As this overview is concerned with the *comparison* of models, the absolute values play a minor role. Tab. 2.1 summarises values that were taken from literature or guessed to the best of the author's knowledge. All parameters were assumed uncorrelated and normally distributed[9], which is, admittedly, a simplification. Cohesion c was omitted because it is assumed to be zero for dry sand.

[8] As $\mu_f \approx f(\boldsymbol{\mu})$ is constant, it does not contribute to the variance of f.

[9] On the influence of parameter distributions on geotechnical calculations, see e.g. *Oberguggenberger* and *Fellin* [95].

Example

The procedure is shortly illustrated with the empirical formula by *Ruse* and *Vermeer*. They expressed N_D as function of φ only (if cohesion is neglected),

$$\begin{aligned} N_D &= \frac{1}{9\tan\varphi} - 0.05 \quad , \\ \rightsquigarrow \mu_{N_D} &= N_D(\mu_\varphi = 32^\circ) = 0.1278 \quad . \end{aligned}$$

In this simple case, the sensitivity k_φ can be calculated explicitly:

$$\begin{aligned} k_\varphi &= \left.\frac{\partial N_D}{\partial \varphi}\right|_{\varphi=32^\circ} \frac{32^\circ}{\mu_{N_D}} \\ &= -\frac{1}{9\sin^2(32^\circ)} \cdot \frac{32^\circ \cdot (\pi/180^\circ)}{0.1278} \\ &= -1.7290 \quad . \end{aligned}$$

Also the variance, and thus the standard deviation for N_D, can be determined

$$\begin{aligned} V_{N_D}^2 &= (k_\varphi V_\varphi)^2 \\ &= (-1.7290 \cdot 0.10)^2 \\ &= 0.0299 \quad , \\ \rightsquigarrow \sigma_{N_D} &= \sqrt{V_{N_D}^2} \cdot \mu_{N_D} \\ &= \sqrt{0.0299} \cdot 0.1278 = 0.0221 \quad . \end{aligned}$$

The dimensionless sensitivities require calculation of partial derivatives of f at point $\boldsymbol{\mu}$. Only for simple functions this can be achieved analytically. If f is more complicated (and thus analytical differentiation error-prone) or not given in explicit form, one must resort to (external) numerical differentiation (cf. e.g. *Ostermann* [98]).

Comparison of the models

Five models (one *Horn* model[10], *Krause*, *Léca/Dormieux Kolymbas* and *Ruse/Vermeer*) were investigated in above described manner with respect to dimensionless sensitivities k_i and uncertainty of the result σ_{N_D}.

Fig. 2.34 illustrates the variation of the result for a variation in input parameters. The bars indicate the approximate 95% confidence intervals based on the calculated

[10] The following configuration was chosen: $K_{\text{silo}} = 0.8$, $K_{\text{wedge}} = 0.4$, distribution of lateral earth pressure according to *DIN 4126*, *no* friction on top of the wedge, approximation of the tunnel face as a square with side length D. This configuration was suggested by *Anagnostou* and *Kovári* [3].

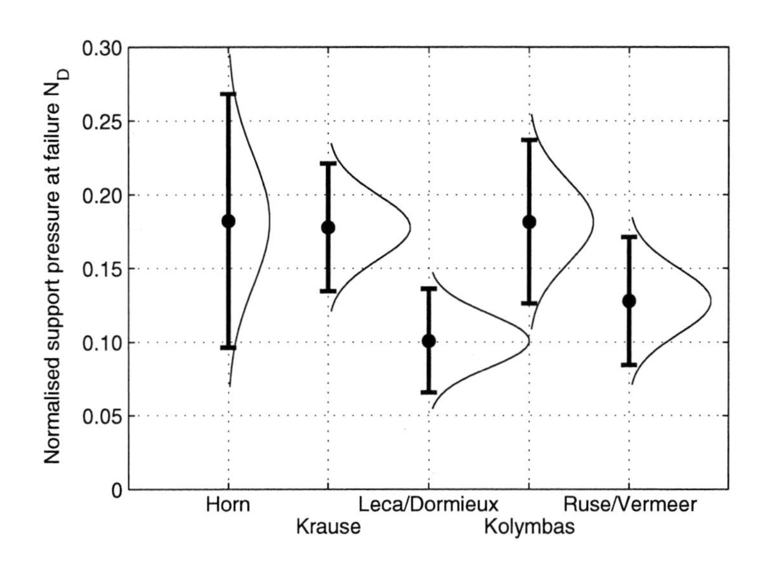

Figure 2.34: Variation of predicted N_D for distributed input data

	Horn	***Krause***	***Léca/ Dormieux***	***Kolymbas***	***Ruse/ Vermeer***
μ_{N_D}	0.1823	0.1778	0.1009	0.1815	0.1278
σ_{N_D}	0.0438	0.0221	0.0180	0.0282	0.0221
k_φ	-2.4051	-1.2428	-1.7833	-1.5562	-1.7290
k_γ	0	0	0	0	0
k_D	-0.0661	0	0	-0.1815	0
k_C	0.0661	0	0	0.1815	0

Table 2.2: Results of the sensitivity study

model variance. The probability distribution functions are also plotted, assuming a normal distribution of N_D. Apart from the different predictions μ_{N_D} (points), it becomes obvious that the different models process the input uncertainties differently: e.g. the distribution of N_D for the *Kolymbas* model is much narrower than that for the *Horn* model, although they predict the same μ_{N_D}.

The individual influence of the input parameters on the prediction for N_D is illustrated in Tab. 2.2. The results show that the friction angle φ is the most decisive parameter for all models. Moreover, φ has a negative sensitivity, i.e. a reduction of φ leads to an increase in N_D. The geometric parameters C and D have less influence on the result, the role of self-weight γ is negligible.

It must be kept in mind that the results of this sensitivity analysis depend on the choice of point $\boldsymbol{\mu}$. Moreover, the neglect of higher order terms in the Taylor-series expansion is only allowed for small V_x. But this short overview underlines the strength of sensitivity analysis: it gives an idea about the uncertainty of the model answer. Even if the true standard deviations for the input parameters are not known, the models can be evaluated qualitatively.

Besides, recommendations by *STUVA*, the German *Research Association for Underground Transportation Facilities* [118], explicitly ask for a parametric study for the face stability analysis. A sensitivity study provides even more information.

2.6 Engineering practice

The preceding comparison has shown that there is much room for interpretation and results can easily be on the unsafe side or extremely conservative. So, how does the engineer in charge estimate the necessary support pressures for a tunnel project? How can the engineer on site interpret, if the prescribed support pressure leads to unforeseen ground movements?

This section gathers some available information on current engineering practice of face stability analysis.

Examples from Europe

The preferred methods for face stability analysis used by German contractors are compiled in Tab. 2.3.

Project	country	D (m)	Applied model
Weser tunnel, Lower Saxony	GER	11.65	DIN 4085
City Tunnel, Leipzig	GER	9.00	Horn
Tiergarten tunnel, Berlin	GER	9.00	DIN 4085
Urban light railway, Cologne	GER	6.52	Jancsecz
	GER	6.85	DIN 4085
Elbe tunnel, 4th tube, Hamburg	GER	14.20	Jancsecz
Herrentunnel, Lübeck	GER	11.80	Jancsecz, DIN 4085
Subway tunnels, Duisburg	GER	6.57	Anagnostou/Kovári
Nodo di Bologna	ITA	9.40	Anagnostou/Kovári
San Vito	ITA	12.30	Anagnostou/Kovári
Railway tunnel, Zürich Thalwil	SUI	12.30	Anagnostou/Kovári
Leidingen tunnel, Rotterdam	NED	5.14	Anagnostou/Kovári
Botlek tunnel, Rotterdam	NED	9.70	Jancsecz

Table 2.3: Examples of applied face stability models for tunnelling projects in Europe (GER: Germany, ITA: Italy, SUI: Switzerland, NED: The Netherlands), [8, 66, 110, 118]

According to Tab. 2.3, three models are currently in use:

- 3D wedge model by *Anagnostou* and *Kovári* [3],
- 3D wedge model by *Jancsecz* [60],
- German Code of Practice *DIN 4085* for earth pressure calculations [34].

Both wedge models originate from *Horn*'s kinematic model; they only differ in their configuration, as explained in Sec. 2.1.1. *DIN 4085* standardises calculation methods for the spatial earth pressure distribution, e.g. on retaining structures. The same methods and assumptions are transferred to face stability calculations.

As mentioned in Chap. 1, the earth pressure at tunnel crown and invert is checked for different modes of operation (cf. e.g. *STUVA* [118] or *Gabener* et al. [42]). Therefore, approaches that include some assumptions of the earth pressure distribution on the face are particularly attractive.

Examples from Japan

Kanayasu et al. [63] reported applied face pressures in Japanese tunnelling projects, for both EPB and slurry shield machines (Tab. 2.4). For EPB applications the necessary support pressure was often calculated as earth pressure at rest, with additional water pressure and a safety margin depending on the ground conditions. In case of slurry shields the support pressure was mainly calculated as water pressure plus, in some cases, active earth pressure and safety margin.

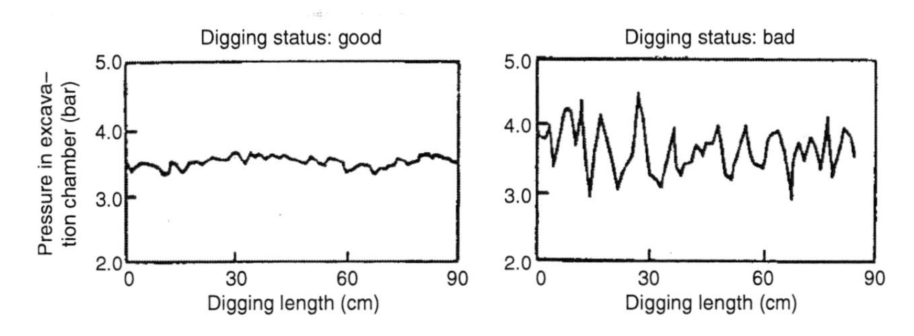

Figure 2.35: Example for possible pressure fluctuations at the tunnel face, [63]

The quoted safety margin was supposed to allow for pressure fluctuations during the driving process. Measurements from an EPB shield heading in Tokyo (Fig. 2.35) show that for a "bad digging status" [63] pressure fluctuations up to 1 bar occurred, indicating a poor state of face stability.

D (m)	Soil type	Applied support pressure
	EPB-shields	
7.45	soft silt	earth pressure at rest
8.21	sandy soil, cohesive soil	earth pressure at rest + water pressure + 0.2 bar
5.54	fine sand	earth pressure at rest + water pressure + fluctuating pressure
4.93	sandy soil, cohesive soil	earth pressure at rest + 0.3-0.5 bar
2.48	gravel, bedrock, cohesive soil	earth pressure at rest + water pressure
7.78	gravel, cohesive soil	active earth pressure + water pressure
7.35	soft silt	earth pressure at rest + 0.1 bar
5.86	soft cohesive soil	earth pressure at rest + 0.2 bar
	Slurry shields	
6.63	gravel	water pressure + 0.1-0.2 bar
7.04	cohesive soil	earth pressure at rest
6.84	soft cohesive soil, diluvial sandy soil	active earth pressure + water pressure + 0.2 bar
7.45	sandy soil, cohesive soil, gravel	water pressure + 0.3 bar
10.00	sandy soil, cohesive soil, gravel	water pressure + 0.4-0.8 bar
7.45	sandy soil	loose earth pressure + water pressure + fluctuating pressure
10.58	sandy soil, cohesive soil	active earth pressure + water pressure + 0.2 bar
7.25	sandy soil, gravel, soft cohesive soil	water pressure + 0.3 bar

Table 2.4: Examples of applied face pressures for tunnelling projects in Japan, [63]

Ultimate and serviceability limit states

According to the principles of limit state design, the ultimate and the serviceability limit states have to be distinguished. For the given problem, the ultimate limit state refers to the collapse of the tunnel face; the serviceability limit state is characterised by displacements at the face and/or at the ground surface that are still acceptable.

From a practical point of view, it is equally important to limit surface settlements to acceptable amounts as to prevent a collapse of the tunnel face. Virtually all theoretical models (cf. Sec. 2.1) focus on the latter, strains/displacements are not or cannot be considered.

Numerical models are able to predict both the limit load and corresponding displacements in the ground. As indicated in Sec. 2.3, the application of these methods is not straightforward; the calculated displacements depend on the chosen material model and the input parameters. *Schubert* and *Schweiger* [109] argued for more sophisticated material models, which are capable of modelling hardening or softening of the material.

Concluding this section: guidelines for the calculation of necessary support pressures, applicable design loads and a sound safety concept are, virtually, still missing for engineering practice. An attempt to standardise existing methods of analysis was made by *STUVA* [118]. Still, the authors pointed out that the quality of a theoretical approach can only be assessed by in-situ measurements or, as those are not available, model experiments.

SYNOPSIS

The face stability problem has been approached in a large number of theoretical, experimental and numerical investigations. A comparison of some proposed models has shown that the predictions for a simple example calculation show a large amount of scatter. Also the propagation of uncertain input parameters is quite different for the models in question.

Despite of the large choice of models, only a few are applied in engineering practice. So far, there is only limited evidence for the quality of these models.

Chapter 3

Experimental investigation of tunnel face stability

Physical modelling has a long tradition in geotechnical research. As large scale tests are time-consuming, expensive and difficult to interpret (due to variability of ground conditions, layering, inhomogeneity), small scale models have been used. They serve to study the behaviour of soil, its interaction with engineering structures or construction processes. Moreover, they can be used to validate numerical models. *Laudahn* [76] listed various applications of small-scale model tests to geotechnical problems, such as foundations, retaining walls, piles, slope stability, anchors and shallow tunnels.

Basically, two types of small scale model tests can be distinguished: tests under 1g-conditions and centrifuge tests under ng-conditions. The major advantages and disadvantages of both types were already mentioned in Sec. 2.2.1.

To investigate the behaviour of a tunnel face close to failure, the author performed 1g model tests. A parametric study was carried out, varying the density of the soil and the overburden above the tunnel. In a comprehensive experimental study the soil displacements and the necessary support force were investigated.

When resorting to small scale models, special attention must be drawn to scaling laws and scaling effects. These are discussed in Sec. 3.1. For similarity reasons it is very difficult to model cohesive soils with 1g model tests. Therefore, all tests were performed with dry sand.

The investigation of soil movements up to the failure of a tunnel face was performed by means of Particle Image Velocimetry (PIV) and is described in Sec. 3.2. The quantitative investigation of the related support force is outlined in Sec. 3.3.

The stress level encountered in the small-scale experiments was in the order of a few kilopascals. Sec. 3.4 focuses on material properties at such low stress levels. This investigation was part of the experimental study and served as preparation for the numerical investigation (Chap. 4).

A rather philosophical approach by *Lee* [78] can be transferred to the author's tests: *Lee* distinguished between *modelling* (constitutive, physical or numerical) and *test-*

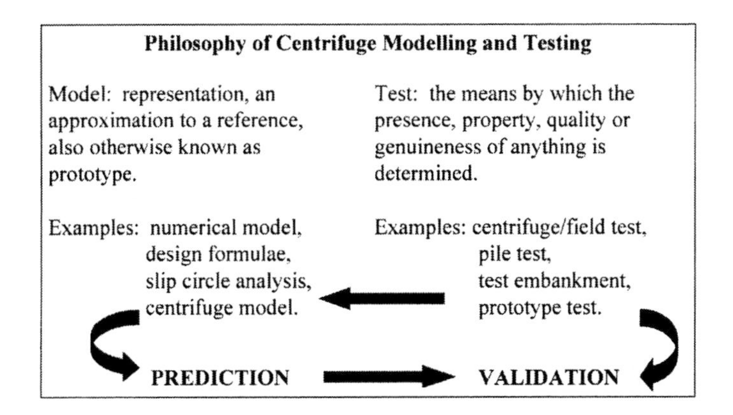

Figure 3.1: Illustration of *Lee*'s concept of modelling and testing, [78]

ing (Fig. 3.1). He argued that "the objective of modelling is to obtain a prediction, whereas that of testing is to obtain a validation" [78]. A test is therefore the complement to a model. According to this definition, both test series can be considered as tests to validate both the kinematic hypotheses of proposed models (first test series, Sec. 3.2) and the predicted necessary forces to ensure face stability (second test series, Sec. 3.3).

In addition, results from both test series are put into context at the end of the section.

3.1 Scaling laws

Dimensional analysis

Physical modelling can only reveal meaningful interpretations, when the measured quantities in the model (index m) are *similar* to quantities on the prototype scale (index p). Basic requirements for the transfer of results from one scale to the other are formulated as scaling laws. These laws derive from dimensional analysis of physical properties. A general overview with mathematical derivations was given by *Görtler* [46]; *Muir Wood* [92] concentrated on problems with geotechnical relevance.

Often, the Π-*theorem* is used to relate results from model and prototype. The theorem states that i governing quantities of a physical problem, x_i, can be expressed in terms of j dimensionless variables, Π_j, which are products of different powers of the x_i. Thus, any functional relation between the physical quantities can be transferred to a relation of dimensionless products (cf. e.g. *Görtler* [46], *Kolymbas* [74]).[1]

[1]It should be underlined that, strictly speaking, the Π_j cannot be created by dividing a physical quantity by an arbitrary reference value. To give an example, division of the stress p by a reference stress p_0 does not give a correct Π-variable, because p_0 can be chosen arbitrarily.

Functional relations between the dimensionless products hold on different scales, if similarity is fulfilled. This is the case for

$$\Pi_{j,\text{model}} = \Pi_{j,\text{prototype}} \quad . \tag{3.1}$$

True similarity is only given if *all* determining properties are scaled correctly with respect to the fundamental dimensions. If not all properties are scaled consistently, similarity is only partly fulfilled. In this case, it remains engineering judgement to decide whether this has a significant influence on the results or not.

Further thoughts on the choice of dimensionless variables for the given problem will be presented in connection with the quantitative experimental study in Sec. 3.3.

Purpose of small-scale modelling

In this study two series of experiments were performed, both with a parametric study for the density index I_d and cover-to-diameter ratio C/D.

The density index is defined as a function of current void ratio e:

$$I_d := \frac{e_c - e}{e_c - e_d} \tag{3.2}$$

with critical and minimum void ratios e_c and e_d. For the low stress levels prevailing in the sandbox experiments,

$$I_d \approx \frac{e_{\max} - e}{e_{\max} - e_{\min}} \quad , \tag{3.3}$$

$e_{\max}$ and $e_{\min}$ being the maximum and minimum void ratios determined from standardised density index tests (e.g. *DIN 18126* [33]).

The purpose of the first series was to reveal realistic failure mechanisms, which could then be put into context with the theoretical models presented in Chap. 2. Theoretical models are usually scale-independent and can, therefore, be equally used for problems on the model and prototype scale.

In the second series of experiments forces were measured. In this case a quantitative transfer of results from the model to prototype scale is not easily possible, even with dimensionless Π-products. The reasons will be outlined in the following paragraphs and in Sec. 3.4. Still, a quantitative comparison of different geometric configurations or soil states *on the model scale* is possible.

Scaling effects

In small scale model tests usually all geometric properties of the prototype are scaled to the same amount, e.g. tunnel diameter $D_{\text{prototype}}$ divided by (geometric) scaling factor n = model tunnel diameter D_{model}, etc. But in most cases, the tested material is *not* scaled according to the scaling rules, which is referred to as *scaling effect* [76]. It is, on the contrary, common practice to use the prototype sand (characterised, e.g., by its mean diameter d_{50})[2] for the model as well. Thus,

$$\left(\frac{d_{50}}{D}\right)_{\text{model}} \neq \left(\frac{d_{50}}{D}\right)_{\text{prototype}} \quad . \tag{3.4}$$

In this case, modelling errors can be expected when shear bands form and dilation of the material leads to restraints in the soil body (e.g. *Walz* [134, 135], *Graf* [47], *Technical Committee TC2* [124]). Dilation effects in the model can be largely overestimated with respect to the prototype. Also the relation between mean grain diameter d_{50} and width of the shear band d_s plays a role: *Nübel* [93] found values for d_s/d_{50} between 10 and 20. It is generally assumed that d_s depends on I_d. For example, *Gudehus* [48] mentioned that the shear zone thickness can be as thin as $5 \cdot d_{50}$ for a very dense soil, and can theoretically be of infinite thickness for very loose soil.

Considering above statements, for the same material with the same density index, the width of a shear band in the model and in the prototype would be roughly the same. Thus, the Π-relation (3.4) can be paraphrased to:

$$\left(\frac{d_s}{D}\right)_{\text{model}} \neq \left(\frac{d_s}{D}\right)_{\text{prototype}} \quad . \tag{3.5}$$

An additional scaling effect needs to be considered, before measurements on the model scale can be interpreted quantitatively. One possibility to quantify the scaling effect is the *modelling of models*, i.e. the investigation of a given problem on several model scales with the same sand.

If shear bands do not govern the system behaviour, some measures can be taken to minimise the scaling effect: the *Technical Committee TC2* (Physical Modelling in Geotechnics) of the ISSMGE put together a catalogue of scaling laws and similitude questions in centrifuge modelling [124]: to minimise grain size effects on soil-structure interaction of a tunnel face stability problem, the following relation is recommended (cf. also *Chambon* et al. [26]):

$$\frac{D}{d_{50}} > 175 \quad . \tag{3.6}$$

[2] The characterisation of the geometric properties of the sand with d_{50} is arbitrary. Other parameters such as coefficient of uniformity U or grain shape influence the material behaviour as well.

In the experiments for the present work two model sands (cf. Sec. 3.2) were investigated, a sand $S1$ with a mean diameter d_{50} = 0.58 cm ($\rightsquigarrow D/d_{50} \approx 170$) and a fine sand $S2$ with d_{50} = 0.24 cm ($\rightsquigarrow D/d_{50} \approx 420$). Thus, (3.6) was more or less fulfilled.

Soil non-linearity

Muir Wood [92] underlined the importance of taking soil non-linearity into account: even if a set of dimensionless variables is found that characterises the given problem, it must be thoroughly investigated if these variables are really scale-independent: to give an example, the friction angle φ can be considered a dimensionless parameter in physical modelling of soils; the Π-theorem states that $\varphi_{\mathrm{model}} = \varphi_{\mathrm{prototype}}$, which does not hold generally for sand in 1g model tests (cf. Sec. 3.4). Also the soil stiffness needs to be treated with care, if the displacements of the soil body are of primary concern.

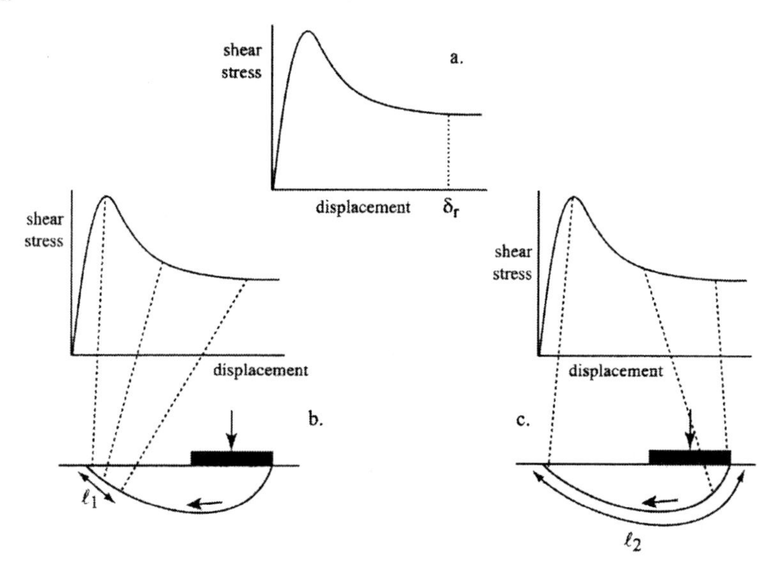

Figure 3.2: Difference in prototype (b) and model (c) behaviour for strain softening material (a), [92]

Further difficulties arise for problems with relative movements along interfaces, either between blocks of soil or between soil and structural elements (*Muir Wood* [92]): if the mobilised shear stress on such an interface depends on relative displacements, the system response on model and prototype scale might be considerably different. This is especially important when the soil shows strain softening behaviour (Fig. 3.2 a).

To elucidate the point, consider the footing shown in Fig. 3.2; due to the reduced scale, a small model footing can only mobilise "small" relative displacements along a possible shear plane (Fig. 3.2 c). Therefore, relatively high shear stresses can be mobilised in comparison with the prototype situation (Fig. 3.2 b), where the shear stresses are at their residual value over almost the entire length of the shear plane; this leads, for the footing example, to an overestimation of the collapse load with a small scale model test.

As a possible remedy, *Muir Wood* [92] suggested to normalise displacements with (mean) particle diameter; this idea will be further illustrated in Sec. 3.2.5.

Above remarks should be kept in mind when looking at the results of both test series. Further reference to the scaling laws will be made where necessary.

3.2 Qualitative investigation of face stability

The first series of experiments aimed at a better understanding of the evolution of failure mechanisms in front of the tunnel face. For this purpose, the soil displacements in the vertical plane of symmetry were followed throughout each test.

As two main parameters the density index of the soil and the overburden were varied. Two sands with different grain-size distributions were used to investigate the scaling effect.

3.2.1 Experimental setup and tested materials

Sandbox and tunnel model

The first series of experiments was conducted in a model box (Figs. 3.3 and 3.4) with inner dimensions 37.2×28.0×41.0 (width × depth × height in cm). The outer frame was made of steel, bottom and side walls wooden, and the front wall was 1 cm thick hardened glass. Thus, the soil grains adjacent to the glass wall could be observed throughout the test.

The problem was modelled in half, cutting vertically through the tunnel axis. Therefore, the tunnel was represented by a half-cylinder of perspex, with an inner diameter of 10.0 cm and a wall thickness of 0.4 cm. This model tunnel protruded 7.0 cm into the soil domain, and its axis lay approx. 8.0 cm above the bottom of the sandbox. An aluminium piston was fitted into the tunnel to support the soil.

The piston was mounted on a horizontal steel rod, and its perimeter was covered with a felt lining. The felt lining prevented sand grains from entering the gap between piston and glass wall. Also, the risk of scratching the glass in course of the experiment was reduced.

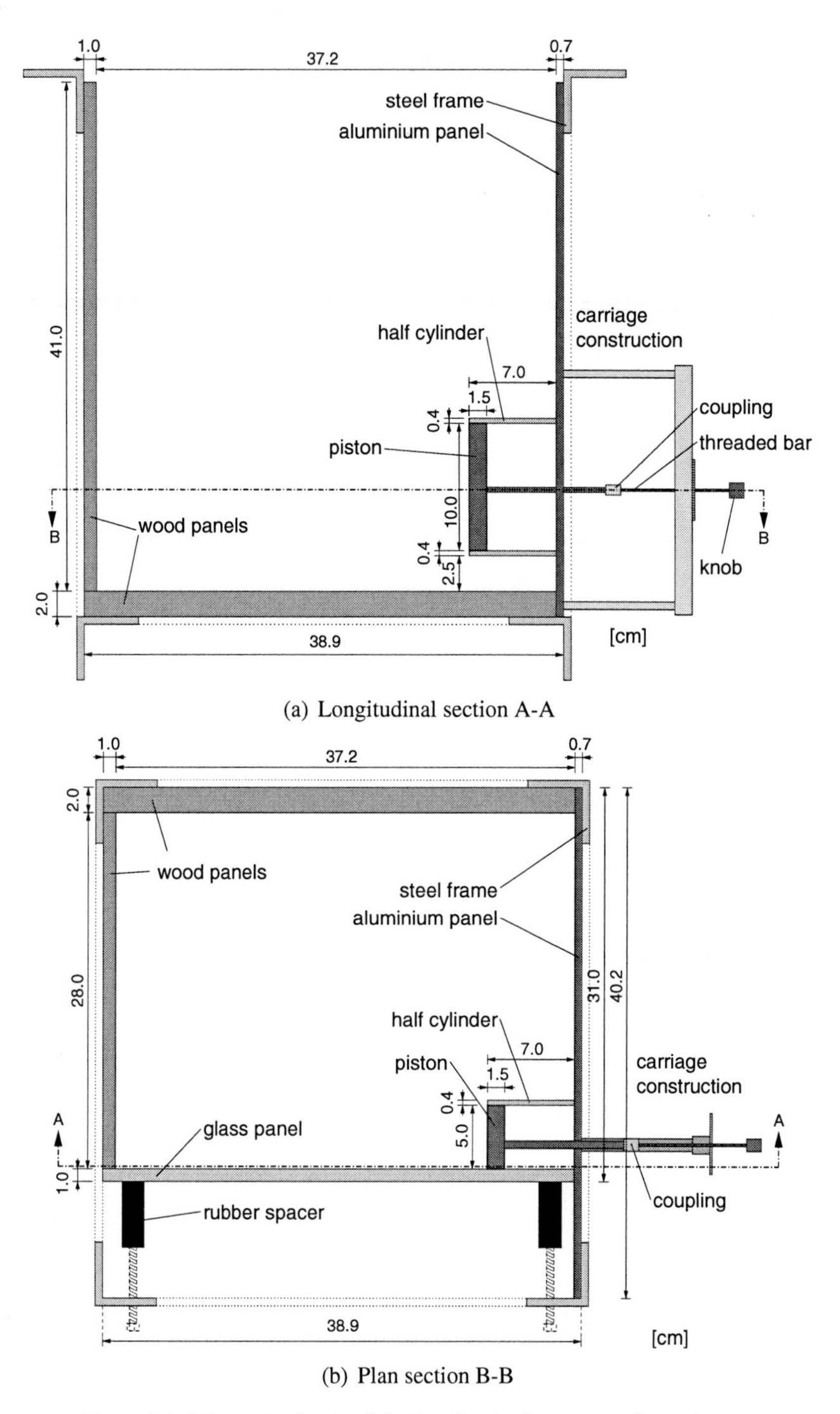

(a) Longitudinal section A-A

(b) Plan section B-B

Figure 3.3: Schematic sketch of the box for the first series of experiments

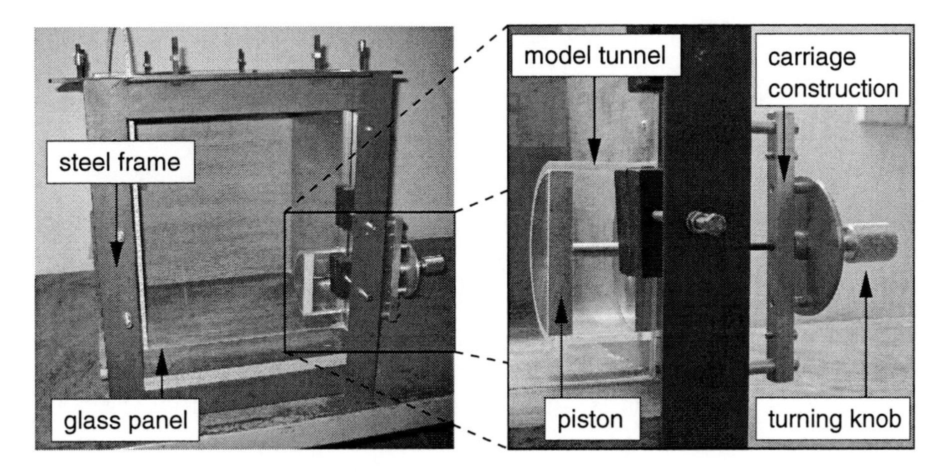

Figure 3.4: Box and model tunnel for the first series of experiments

In the first series of experiments no forces were measured; therefore, it was not considered necessary to reduce the friction between piston and tunnel wall.

The rod was supported by a one-dimensional roller bearing inside the side wall of the box that allowed for horizontal movements in direction of the tunnel axis. To trigger collapse of the face, the piston could be retracted into the model tunnel by turning a threaded bar and a knob in extension of the piston axis (Fig. 3.4, right). One revolution of the knob led to 1 mm horizontal displacement.

Sand properties

As in preceding studies (*Laudahn* [76], *Mähr* [81]), commercially available *Siligran* quartz sand from the deposit Ottendorf-Okrilla in Saxony (Germany) was used.[3]

For the model tests two different grain size distributions were used (Fig. 3.5 a): a sand with grain diameters between 0.1 and 2.0 mm ($S1$) and a fine sand with grain sizes between 0.1 and 0.5 mm ($S2$). Fig. 3.5 b shows the grain shapes; more properties are listed in Tab. 3.1.

3.2.2 Test procedure

Preparation of the sand body

The tests were performed with dry sand for various C/D ratios and different initial densities I_d. The densities were not defined a priori, because the focus of this investigation was on *qualitative* differences in the beheviour of dense and loose samples.

[3]The sand is distributed by the company EUROQUARZ GmbH, D-01936 Laußnitz, Würschnitzer Str. 2. Before delivery it is washed, dried and sieved.

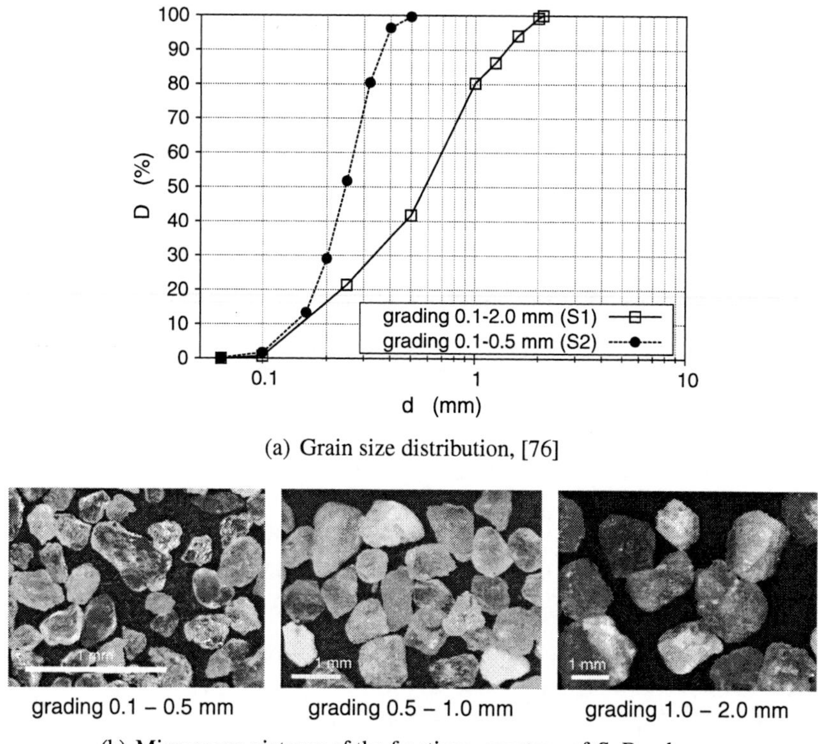

(a) Grain size distribution, [76]

(b) Microscope pictures of the fractions, courtesy of *S. Berghamer*

Figure 3.5: Characteristics of the applied sand

Property		**Sand (*S1*)**	**Fine sand (*S2*)**
Mean grain size	d_{50}	0.58 mm	0.24 mm
Coefficient of uniformity	U	3.6	1.9
Grain shape		angular to subangular	
Specific weight	ρ_s	2.635 g/cm^3	
Max. void ratio	$e_{\max}$	0.75	0.87
Min. void ratio	$e_{\min}$	0.42	0.56

Table 3.1: Properties of applied Ottendorf-Okrilla sands, cf. [76]

It turned out that the resulting I_d values fell within a reasonably small range, if the following preparation procedures were pursued:

The *dense* samples were prepared by dry pluviation of sand into the box with a funnel. With a drop height of approx. 10 cm, layers of 5 cm thickness were installed. These layers were then compacted manually by hand tamping[4] with approximately the same compaction energy for each layer.

[4] A tamper with a square base of approx. 15 × 15 cm was used.

The *loose* samples were prepared by carefully putting the sand into the box with a small shovel, trying to prevent any compacting action.

The quoted void ratios e and resulting densities I_d must be understood as average values throughout the whole sample. For each test the void ratio e was calculated as

$$e = \frac{\rho_s}{m_d/V} - 1 \quad ,$$

with the mass of the sand inside the box m_d and the occupied volume V. I_d was then determined with (3.3).

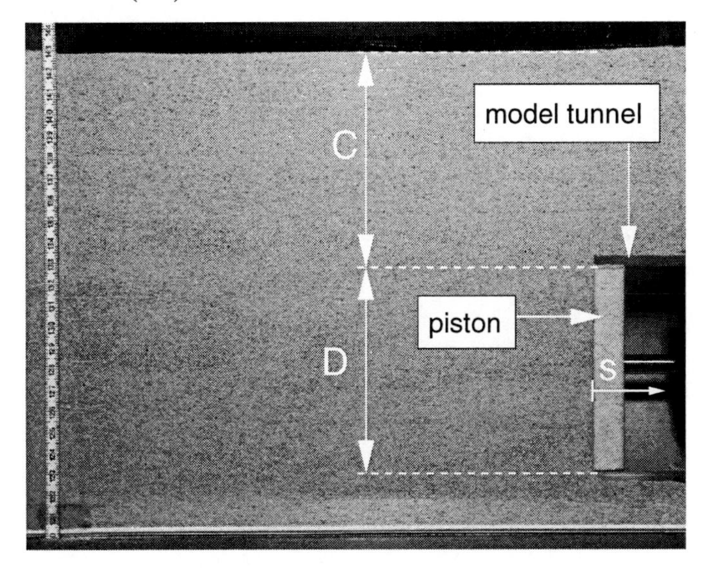

Figure 3.6: Soil sample and model tunnel at the start of the test

During sample preparation the piston was fixed at the front of the model tunnel, thus preventing any soil from entering the cylinder. Fig. 3.6 shows a picture of the box before the start of a test.

Steps of the model tests

In the course of the experiment the piston was retracted into the model tunnel, which triggered the face collapse. In doing so, the soil *displacements* at the tunnel face were prescribed, but not the (support) *pressure*. This was, admittedly, not in agreement with the real problem, where the *pressure* inside the slurry is adjusted. Other researchers, such as *Chambon* and *Corté* [24] or *Mélix* [88], used inflatable membranes to support the soil. But none of them modelled half the problem to investigate the evolution of a possible failure mechanism. In case of cohesionless soil the application of a membrane was not considered practicable. Moreover, it is not possible to capture strain softening effects of the material in a stress-controlled test. Thus, the author decided to prescribe the displacements of the soil in front of the face, a solution also adopted by e.g. *Kamata* and *Mashimo* [62].

For the first 6.0 mm of piston displacement, increments Δs =0.25 mm were chosen, which were applied by turning the knob by 45°. This increment size corresponds to 0.5 $d_{50,S1}$ and 1.0 $d_{50,S2}$ and was considered small enough to capture the soil deformation at the onset of failure with sufficient accuracy. For another 19 mm, Δs was 0.5 mm. In doing so, an overall displacement of 25 mm took place, corresponding to 25 % of the tunnel's inner diameter.

After each increment a digital picture of the grain structure was taken (cf. Sec. 3.2.3).

Test programme

After a number of preliminary tests to optimise the experimental setup, a total of 12 tests were performed[5]; their configurations are summarised in Tab. 3.2.

Test	C/D	material	density	e_0	I_d
1	0.5	$S1$	dense	0.49	0.80
2	0.5	$S2$	dense	0.57	0.98
3	0.5	$S1$	loose	0.66	0.27
4	0.5	$S2$	loose	0.81	0.20
5	1.0	$S1$	dense	0.48	0.83
6	1.0	$S2$	dense	0.60	0.86
7	1.0	$S1$	loose	0.66	0.26
8	1.0	$S2$	loose	0.81	0.18
9	1.0	$S1$	dense	0.47	0.85
10	1.0	$S2$	dense	0.59	0.91
11	1.0	$S1$	loose	0.66	0.29
12	1.0	$S2$	loose	0.83	0.13

Table 3.2: Test programme for the PIV investigation

[5]The experiments were performed by *Ivan Doro* [36], *Claudio Ravazzini* [105] and *Thomas Ashworth* [6] under the supervision of the author.

Particle Image Velocimetry (PIV), synonymous with **Digital Image Correlation (DIC)**, is a non-invasive technique that allows quantitative investigation of plane displacement patterns. It was first applied in fluid dynamics to visualise flow-fields, but is nowadays applied in various research areas, such as heat propagation or aerodynamics [50].
In the last years it has been increasingly applied to geotechnical applications because it allows to investigate the displacement fields in a soil sample on a grain-scale level (e.g. *Nübel* [93, 94], *White* et al. [136, 137], *Mähr* [81], *Hauser* [50]). Details on the method and its application can be found in *Raffel* et al. [103].

3.2.3 Evaluation of soil displacements with Particle Image Velocimetry

The qualitative investigation served to identify displacement patterns during the collapse of a tunnel face. The model tests for the present work were evaluated with Particle Image Velocimetry. Its advantage over coloured soil layers[6] is that the whole soil domain is under survey, and that the appearance of shear bands can be detected much easier. Moreover, preparation of the sand body is less time consuming.

Input for the PIV analysis were high-resolution pictures of consecutive (displacement) states of the grain skeleton, which were taken through the glass panel.

Equipment and setup

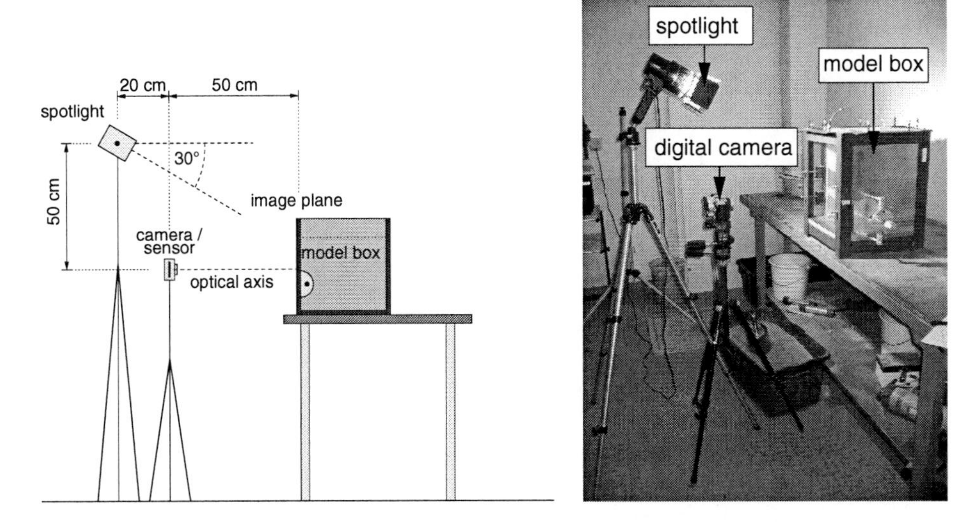

Figure 3.7: Setup of the tests

[6]A simple technique to observe soil displacements is the marking of soil layers with coloured material. The displacements of the whole body can be deduced from the observed changes in shape of the marker lines.

Fig. 3.7 shows the setup of the experiment with sandbox, camera and spotlight. The experiments were executed in a laboratory room without daylight. The spotlight, type OSRAM Video 1000, was the only source of light, supplying a constant illumination of the box throughout a test. In order to create a diffuse lighting regime, a white sheet of paper was fixed on the spotlight. Moreover, to avoid any reflection of the light source on the resulting pictures, the spotlight was positioned 50 cm above the camera, with an approximate inclination of 30° versus the horizontal (Fig. 3.7).

Pictures were taken with a digital camera *Minolta Dimage 7i* with a maximum resolution of 5 Megapixels (2560 px × 1920 px). Not only the camera settings (such as resolution or image compression) were important for the quality of the resulting pictures; also the positions of camera and spotlight, and the orientation of the camera axis with respect to the glass panel had to be adjusted. All pictures had to be taken remote-controlled, because pressing the shutter button of the camera would have changed the camera position too much.

In a preliminary study the setup was optimised (Tab. 3.3):

Camera settings	
Colour mode	black & white
Photo sensitivity	ISO 200
Focal distance	40-50 mm
Geometric setup	
Distance camera - sandbox	50 cm
Distance spotlight - sandbox	70 cm

Table 3.3: Optimised experimental setup for the PIV evaluation, [36]

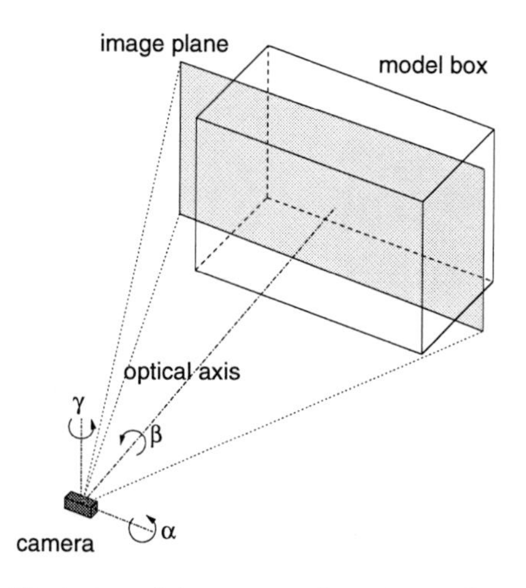

Figure 3.8: Illustration of the camera orientation angles α, β and γ

The camera orientation plays an important role for the quality of the pictures and the subsequent evaluation. Fig. 3.8 shows the three angles α, β and γ that determine the orientation of the camera axis with respect to the glass front of the sandbox (cf. *Rinawi* [106]). Their correct adjustment influences the quality and the comparability of the pictures to a great extent.

- The angle α determines the horizontality of the camera axis. In the correct position the camera axis is horizontal. The alignment was made by means of small water-levels that were mounted on the camera tripod.
- The angle β describes a rotation of the subject around the optical axis of the camera. This angle was also aligned by a small water-level on the tripod. Moreover, it was checked whether horizontal lines of the subject appeared horizontal on the camera display.
- The angle γ describes a rotation about the vertical axis. It is very important to adjust γ correctly, because otherwise equidistant points on the subject do not appear equidistant on the picture. Therefore, a mirror was temporarily attached to the glass front of the box. By making sure that the centre of the camera objective was perfectly centred in the resulting picture on the display, γ was aligned. This also served as a check for the correct alignment of α.

Even though the three angles are theoretically independent, it was an iterative process to adjust them in practice until a satisfying result was achieved. Moreover, it had to be guaranteed that the glass front was vertical and the tunnel axis horizontal for the described measures to be effective.

Postprocessing of the pictures

In the present study only quasi-static deformations were regarded; inertia was not considered a governing model parameter, and the behaviour of dry sand was assumed to be rate-independent. The digital pictures were transferred to a computer after each test, where they were processed by the free PIV package MatPIV v. 1.6.1 by *Sveen* [119]. It consists of several routines for the commercial software MATLAB.

The basic idea of PIV is image correlation of *interrogation cells* in consecutive pictures. These interrogation cells cover a few sand grains and are characterised by a certain distribution of grey or colour values. In this case, the pattern of light and shadow in the grain assembly served as tracer.

Fig. 3.9 illustrates the evaluation procedure: the characteristics of the interrogation cell in picture $i-1$ are compared with characteristics of search cells of the same size in picture i. For small displacements between two subsequent pictures, it can

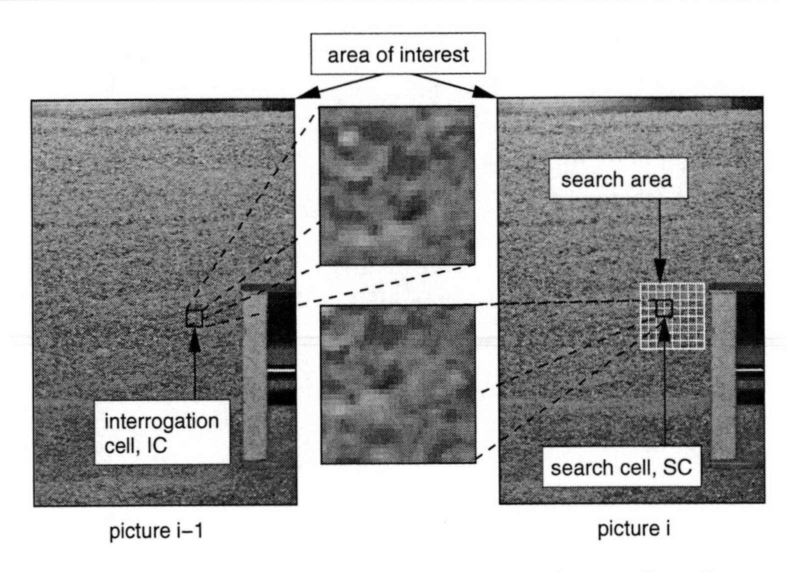

Figure 3.9: Consecutive digital pictures with with interrogation cell

be assumed that the pattern of an interrogation cell remains unchanged, and only the position of the cell varies.

Then a local displacement vector $\mathbf{u}(\mathbf{x}, t_i)$ can be determined by means of the cross-correlation:

$$\mathbf{R}(u_1, u_2) = \sum_{i=-M/2}^{M/2} \sum_{j=-N/2}^{N/2} \mathrm{Im}_1(i,j)\,\mathrm{Im}_2(i+u_1, j+u_2) \quad , \qquad (3.7)$$

$\mathrm{Im}_1(i,j)$ ... light intensity of pixel (i,j) of interrogation cell in picture 1,
$\mathrm{Im}_2(i,j)$ ... light intensity of pixel $(i+u_1, j+u_2)$ of search cell in picture 2,
M ... width of interrogation cell (in pixel),
N ... length of interrogation cell (in pixel),
R ... cross correlation coefficient.

The correlation coefficient $R(u_1, u_2)$ measures the agreement between interrogation and search cell.[7] Fig. 3.10 shows the results of the cross-correlation. The position of its peak R_{max} defines the most probable new position of the interrogation cell in the following picture. R_{max} also serves as quality measure for the detected displacement: the closer R_{max} to one, the better the correlation. The applied PIV algorithm is able to determine the position of the peak with sub-pixel accuracy. This is achieved by interpolation of the discrete correlation values.

In this fashion the whole problem domain is evaluated, resulting in a matrix of incremental displacements $\mathbf{u}(\mathbf{x}, \Delta t_i)$ between pictures $i-1$ and i. It should be noted

[7] As the light intensity distributions are normalised, a value of $R = 1$ indicates absolute agreement between the two cells.

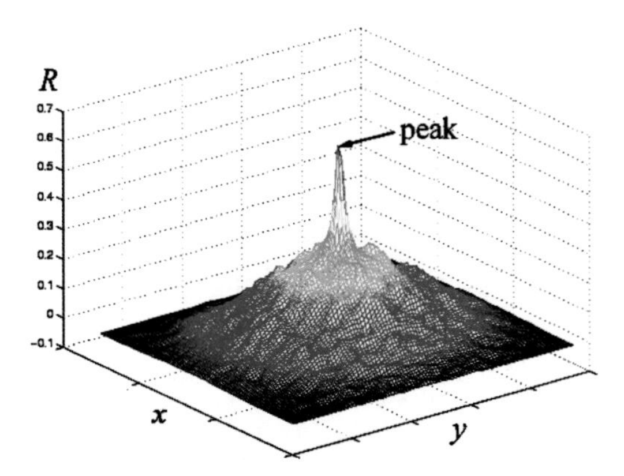

Figure 3.10: Illustration of the cross-correlation matrix for a single interrogation cell, [81]

that the results represent an EULERIAN description of the movement, because the software uses a fixed reference coordinate system for all steps. This was appropriate for the purpose of this study, because the vector fields of incremental displacements $\mathbf{u}(\mathbf{x}, t_i)$ can be interpreted as velocity fields. These were then used for the detection of failure mechanisms.

Figure 3.11: Coordinate system used for relation between picture and world coordinates

To relate the *picture coordinates* (given in pixels) to real *world coordinates* (in e.g. cm) the PIV routine needs reference points on one of the digital pictures. Therefore, a printed coordinate grid was temporarily attached to the glass panel (Fig. 3.11).

The pictures of the sandbox covered an area of 36 cm by 27 cm. With an image size of 2560 px $\times$ 1920 px and a mean grain diameter $d_{50,S1}$ = 0.58 mm of material $S1$, the resolution of the pictures was roughly 0.14 mm/px and 0.25 $d_{50,S1}$/px. For the fine sand the resolution in terms of mean grain diameter was 0.6 $d_{50,S2}$/px.

Visualisation of results

The primary evaluation results of PIV are vector fields of incremental displacements for a given increment Δt from t_{i-1} to t_i:

$$\mathbf{u}(\mathbf{x}, \Delta t_i) = \begin{pmatrix} u_1(x_1, x_2, \Delta t_i) \\ u_2(x_1, x_2, \Delta t_i) \end{pmatrix} = \mathbf{x}(t_i) - \mathbf{x}(t_{i-1}) = \begin{pmatrix} x_1(t_i) - x_1(t_{i-1}) \\ x_2(t_i) - x_2(t_{i-1}) \end{pmatrix} .$$

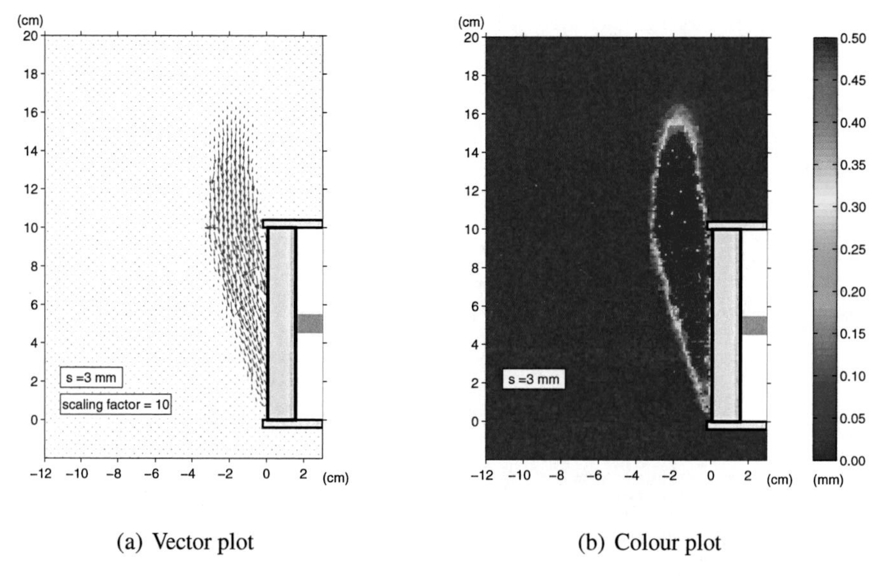

(a) Vector plot (b) Colour plot

Figure 3.12: Visualisation of incremental displacements for a piston advance from 2.25 to 2.50 mm (dense sand, C/D = 1.0)

The fields of incremental displacements can be visualised with *arrows* (Fig. 3.12 a). Thus, the *directions* of the displacements are well recognisable. Fig. 3.12 b shows a *colour* plot: different lengths of the displacement vectors are represented by different colours. In this mode of visualisation it can easily be seen which parts of the soil body move and which not. However, the information about the direction of the movement is lost.

For illustration purposes, in Fig. 3.12 and subsequent figures the tunnel lining and the piston contour are highlighted.

Mesh study

A mesh study served to determine the optimum size for the interrogation cells. It is desirable to make the cells as small as possible to achieve a high resolution of results. But with decreasing cell size it becomes more difficult for the PIV algorithm to relate corresponding cells in two subsequent pictures, which becomes manifest in lower correlation peaks.

Several combinations of subsequent sandbox pictures were evaluated with edge lengths of the (square) interrogation cells of 32 px, 24 px, 16 px and 12 px. The cell size was not decreased any further because *Raffel* et al. [103] recommended a minimum size of the interrogation cell of 5 grains, which was still fulfilled for the smallest cell size (12 px $\approx 3\ d_{50,S1}$).

To capture large displacements with sufficient accuracy, an *adaptive multi-pass process* was applied; this means that the PIV evaluation starts with an initial size of the interrogation cell of 128 $\times$ 128 px, which is reduced in several passes to the desired minimum cell size. In this case, the vector field of one pass serves as reference for the next pass.

To further increase the resolution of the resulting displacements field it is possible to use an overlap of the interrogation cells. For this study an overlap of 50% (of the edge length) was applied. It should be kept in mind that the overlap of cells smoothes the displacements field; two incremental displacement vectors in immediate vicinity are not independent of each other.

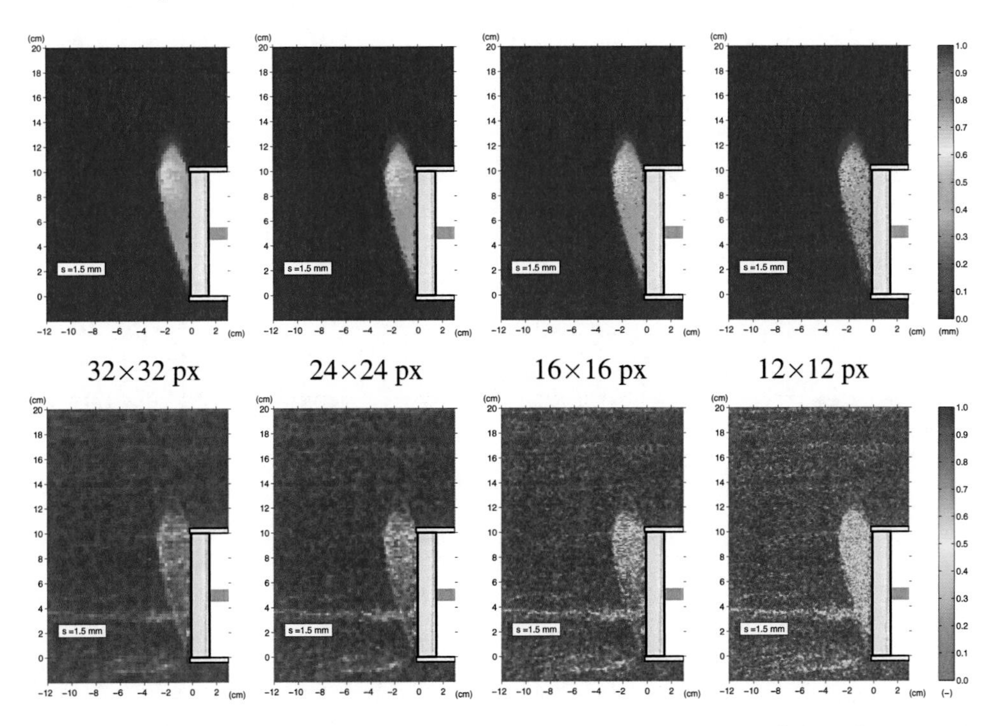

Figure 3.13: Incremental displacements (top row) and maximum correlation coefficients (bottom row) for different sizes of the interrogation cells

Fig. 3.13 (top row) shows colour plots of incremental displacements for a piston movement of $\Delta s = 0.25$ mm. The maximum correlation coefficients R_{max} are also plotted for each grid point (Fig. 3.13, bottom row). It becomes obvious that the shape of the zone in movement is identical for the different cell sizes. But with decreasing cell size the number of grid points with maximum correlation coefficients below, say, 0.5 increases (especially in front of the tunnel). This indicates a decrease in quality of the PIV correlation.

The author decided to perform the main investigation for material $S1$ with a minimum cell size of 16 $\times$ 16 px. This size combines a sufficient quality with a good resolution. Moreover, the edge length corresponds to 4 $d_{50,S1}$, which is smaller than the expected width of shear bands (in the order of 10 to 20 times $d_{50,S1}$ for initially dense samples). For material $S2$ the best results were obtained with a cell size of 24 $\times$ 24 px.

With the applied PIV software it is possible to smooth the vector field, i.e. to detect and remove outliers (e.g. by means of Signal-to-Noise ratio filter or peak height filter, cf. [119]). For all results presented here, no smoothing whatsoever was applied to keep them as "pure" as possible.

Precision of PIV

Before starting the sandbox experiments the measurement uncertainty of the PIV evaluation for the given setup had to be assessed. For this purpose, a sand tile was fixed on the horizontal steel rod inside the model tunnel (Fig. 3.14 a). A movement of 1.0 mm to the right was applied by a full revolution of the knob, and pictures were taken before and after the movement.

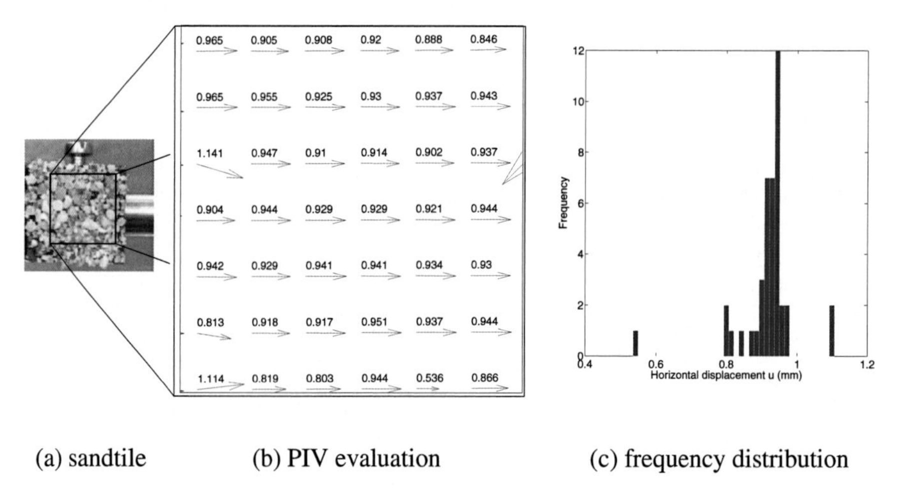

(a) sandtile (b) PIV evaluation (c) frequency distribution

Figure 3.14: PIV evaluation of a sand tile movement of 1 mm

Fig. 3.14 b shows the PIV evaluation of this movement; these PIV evaluations can be interpreted as n measurements, u_i, of the predefined displacement of u = 1.0 mm.

The sample standard deviation s can be approximated by

$$s^2 := \frac{1}{n-1} \sum_n (\bar{u} - u_i)^2 \quad ,$$

$$\rightsquigarrow s = \sqrt{\frac{1}{n-1} \sum_n (\bar{u} - u_i)^2} \quad .$$

with the average $\bar{u}$,

$$\bar{u} = \frac{1}{n} \sum_n u_i \quad .$$

Then the *random* measurement uncertainty p of the measurement method can be approximated by

$$p = s/\sqrt{n} \quad ,$$

In this context, s can be interpreted as mean error of a single measurement u_i, p as mean error of the average $\bar{u}$.

For the horizontal displacement u of the sand tile, the following results were obtained for n = 42 measurements:

$$s_u = 0.08 \text{ mm} \quad ,$$
$$p_u = 0.01 \text{ mm} \quad .$$

Thus, the measurement uncertainty was in the order of one hundredth of a millimetre. The mean horizontal displacement $\bar{u}$ was 0.92 mm. Therefore, the bias, i.e. the *systematic* measurement uncertainty, can be quantified as a_u = 0.08 mm. Errors can be attributed to inaccurate positioning of the camera with respect to the glass panel or imprecise relation between picture and world coordinates.

Measurement uncertainties for PIV evaluations of 0.1 mm are quoted in the literature (*Mähr* [81], *Hauser* [50]). *Nübel* [93] mentioned that this is sufficient to compare experimental observations to results of continuum model simulations, e.g. with finite elements.

Further processing of the information

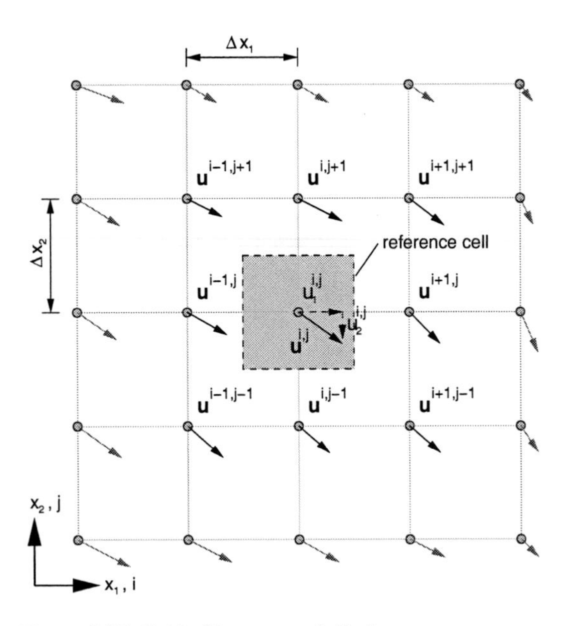

Figure 3.15: Grid of incremental displacement vectors $\mathbf{u}$

With the following assumptions the primary results of the PIV analysis, i.e. the incremental displacement fields, were further processed:

- the soil can be treated as a continuum,
- small strains per increment,
- plane strain conditions.

Incremental strains ε were derived from incremental displacements $\mathbf{u}$ in distinct grid points (Fig. 3.15) making use of a two-dimensional finite difference scheme[8]:

$$\varepsilon = \begin{pmatrix} \varepsilon_{11} & \varepsilon_{12} \\ \varepsilon_{21} & \varepsilon_{22} \end{pmatrix} = \begin{pmatrix} \dfrac{\partial u_1}{\partial x_1} & \dfrac{1}{2}\left(\dfrac{\partial u_1}{\partial x_2} + \dfrac{\partial u_2}{\partial x_1}\right) \\ \dfrac{1}{2}\left(\dfrac{\partial u_1}{\partial x_2} + \dfrac{\partial u_2}{\partial x_1}\right) & \dfrac{\partial u_2}{\partial x_2} \end{pmatrix} , \tag{3.8}$$

[8]Other possibilities are listed in *Raffel* et al. [103].

with

$$
\begin{aligned}
\frac{\partial u_1}{\partial x_1} &\cong \frac{(u_1^{i+1,j+1} + 2u_1^{i+1,j} + u_1^{i+1,j-1}) - (u_1^{i-1,j+1} + 2u_1^{i-1,j} + u_1^{i-1,j-1})}{8\Delta x_1} \quad , \\
\frac{\partial u_1}{\partial x_2} &\cong \frac{(u_1^{i-1,j+1} + 2u_1^{i,j+1} + u_1^{i+1,j+1}) - (u_1^{i-1,j-1} + 2u_1^{i,j-1} + u_1^{i+1,j-1})}{8\Delta x_2} \quad , \\
\frac{\partial u_2}{\partial x_1} &\cong \frac{(u_2^{i+1,j+1} + 2u_2^{i+1,j} + u_2^{i+1,j-1}) - (u_2^{i-1,j+1} + 2u_2^{i-1,j} + u_2^{i-1,j-1})}{8\Delta x_1} \quad , \\
\frac{\partial u_2}{\partial x_2} &\cong \frac{(u_2^{i-1,j+1} + 2u_2^{i,j+1} + u_2^{i+1,j+1}) - (u_2^{i-1,j-1} + 2u_2^{i,j-1} + u_2^{i+1,j-1})}{8\Delta x_2} \quad .
\end{aligned}
$$

Note that, in contrast to an often adapted "geotechnical" definition of strains (i.e. compression positive), the mechanical definition is used here (i.e. compression negative).

The first invariant of the strain tensor, I_1^ε, is a measure for incremental **volumetric strains** (in case of small strains):

$$
I_1^\varepsilon = \mathrm{tr}\,\boldsymbol{\varepsilon} = \varepsilon_{11} + \varepsilon_{22} = \varepsilon_{\mathrm{vol}} \quad . \tag{3.9}
$$

A second invariant of the strain tensor, I_2^ε, the norm of the deviatoric part $\boldsymbol{\varepsilon}^*$, serves as a scalar representation of incremental **shear strains**:

$$
I_2^\varepsilon = \sqrt{\mathrm{tr}\,(\boldsymbol{\varepsilon}^{*2})} = \sqrt{\varepsilon_{ij}^*\,\varepsilon_{ij}^*} \quad , \tag{3.10}
$$

with

$$
\boldsymbol{\varepsilon}^* = \boldsymbol{\varepsilon} - 1/2\,\mathrm{tr}\,\boldsymbol{\varepsilon}\,\mathbf{1} \quad .
$$

The curl of the incremental displacements field serves to highlight the **circulation** of the vector field during the failure process:

$$
\mathrm{curl}\,\mathbf{u} = \frac{\partial u_1}{\partial x_2} - \frac{\partial u_2}{\partial x_1} \quad . \tag{3.11}
$$

Moreover, with $\varepsilon_{\mathrm{vol}}$ it is possible to trace the evolution of **void ratio** e throughout an experiment:

$$
e_i = e_{i-1} + (1 + e_{i-1})\,\varepsilon_{\mathrm{vol},i} \quad . \tag{3.12}
$$

It was pointed out before that the PIV evaluation provides a EULERIAN description of the movement. Relation (3.12) can be derived from the mass balance equation for spatial coordinates (cf. e.g. *Mase* [85]), which will shortly be outlined here:

The positions of the grid points in the PIV evaluation remain fixed with time; these points can, therefore, be interpreted as centres of *reference cells* of the continuum (cf. Fig. 3.15). Volumetric strains are defined at these discrete grid points. As the PIV grid is time-independent, the mass balance equation can be written as

$$\frac{D\rho(\mathbf{x},t)}{Dt} + \rho(\mathbf{x},t)\ \mathrm{div}\ \mathbf{v}(\mathbf{x},t) = 0 \quad , \tag{3.13}$$

with the current density $\rho(\mathbf{x},t) =: \rho$ of the material in a given cell and the velocity field $\mathbf{v}(\mathbf{x},t) =: \mathbf{v}$ for the problem domain.

The change in density with time can be formulated incrementally, in terms of the strains in the given cell:

$$\begin{aligned} \frac{D\rho}{Dt} &= -\rho\left(\frac{\partial v_1}{\partial x_1} + \frac{\partial v_2}{\partial x_2}\right) \quad , \\ \leadsto \Delta\rho &= -\rho\,(\varepsilon_1 + \varepsilon_2) \quad , \\ \leadsto \Delta\rho &= -\rho\,\varepsilon_{\mathrm{vol}} \quad . \end{aligned} \tag{3.14}$$

The relation between change in density, $\Delta\rho$, and change in void ratio, Δe, can be derived as:

$$e = \frac{\rho_s}{\rho} - 1 \quad , \tag{3.15}$$

$$\begin{aligned} \leadsto \frac{\partial e}{\partial t} &= -\frac{\rho_s}{\rho^2}\frac{\partial \rho}{\partial t} \quad , \\ \leadsto \Delta e &= -\frac{\rho_s}{\rho^2}\,\Delta\rho \quad , \end{aligned} \tag{3.16}$$

assuming that the density of the grains, ρ_s is constant.

Making use of (3.14) and (3.15), (3.16) can be formulated as

$$\begin{aligned} \Delta e &= \frac{\rho_s}{\rho}\,\varepsilon_{\mathrm{vol}} \\ &= (1+e)\,\varepsilon_{\mathrm{vol}} \quad , \end{aligned}$$

which finally yields (3.12).

Limitations of PIV for the given problem

Limitations of PIV for the performed experiments are:

- The investigated problem is a three-dimensional one. Obviously, evaluation of pictures of one problem boundary does not reveal the out-of-plane deformation. Nevertheless, expected out-of-plane movements were minimised by choosing the axis of symmetry of the problem for taking the pictures.

- Also friction between sand and glass panel influences the mobility of the sand grains and may, thus, cause a deviation of the observed from the true collapse pattern. This influence was neglected because the focus of the evaluation was not on the *absolute* displacements of the sand grains. More specifically, the directions of movement and relative displacements were important. The author assumed that relative displacements were approximately the same for experiments with and without friction between sand and glass.

3.2.4 Results

The description of PIV results will focus on experiments with material $S1$. If not stated otherwise, the experiments with material $S2$ revealed the same qualitative soil behaviour. An investigation of the scaling effect by comparing materials $S1$ and $S2$ follows in Sec. 3.2.5.

Incremental displacements

Fig. 3.16 shows colour plots of incremental displacements for an advance step from 1.25 to 1.00 mm for different cover-to-diameter ratios and relative densities. The shape and extent of the failure zone did *not* depend on the height of the cover, because the pictures are virtually identical for different C/D values.

The influence of soil density was much more pronounced: while the dense samples showed a clearly defined failure zone, this zone was rather diffuse for the loose samples (Fig. 3.17). Moreover, for the loose samples soil movements reached up to the ground surface, whereas for the dense ones movements were concentrated in the vicinity of the tunnel face.

Development of the failure zone with advancing piston

For the dense samples a propagation of the failure zone was observed. This zone started from a wedge-like structure in front of the piston and subsequently extended vertically upwards.

On the contrary, for the loose sand no distinct development of the failure zone could be detected. From the first two or three advance steps onwards, the shape of the failure zone remained practically identical (equal to the pattern in Fig. 3.16 d-f).

As mentioned above the incremental displacements are the primary result of the PIV evaluation. Other quantities such as strain, void ratio or curl were derived from the displacement field. The differentiation of the displacement field leads to an increase

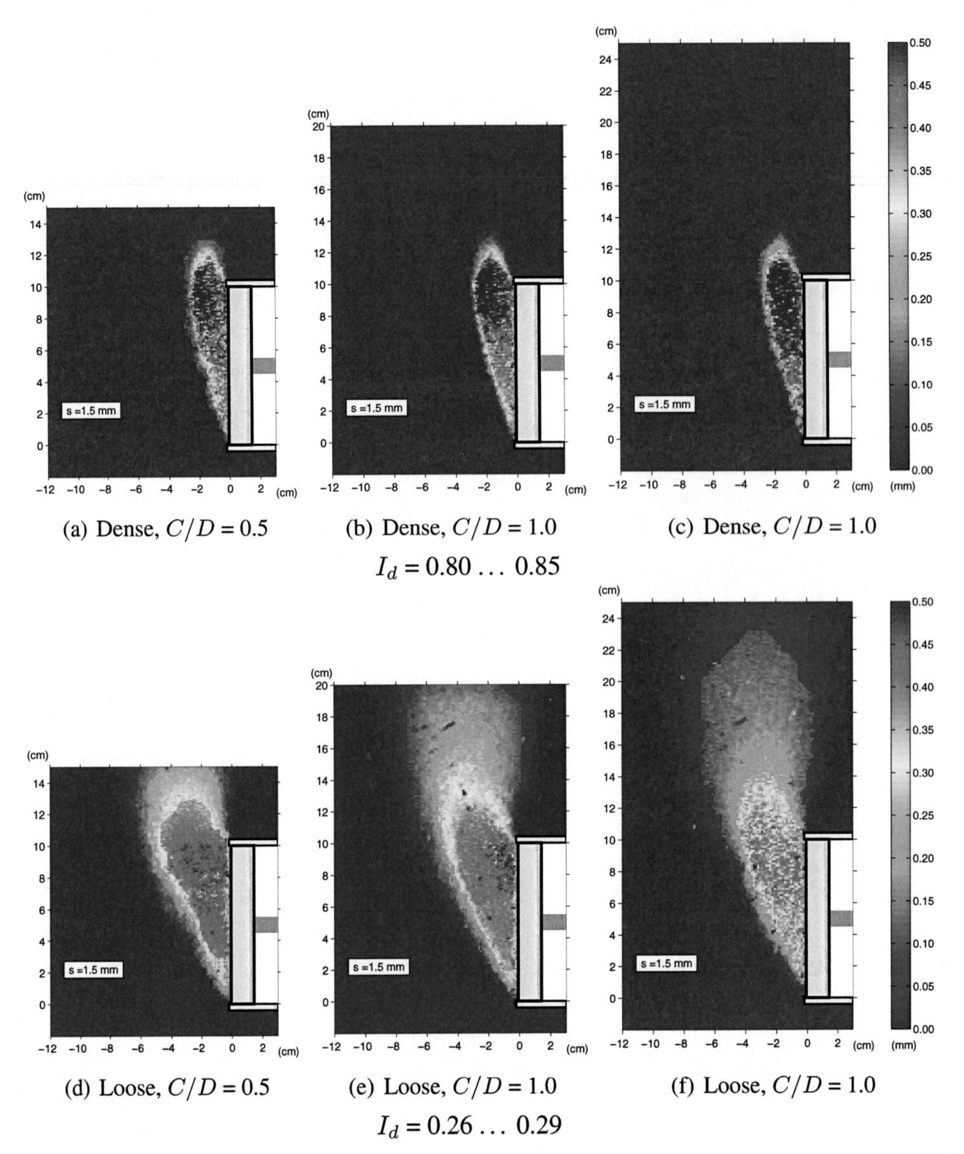

(a) Dense, $C/D = 0.5$ (b) Dense, $C/D = 1.0$ (c) Dense, $C/D = 1.0$

$I_d = 0.80 \ldots 0.85$

(d) Loose, $C/D = 0.5$ (e) Loose, $C/D = 1.0$ (f) Loose, $C/D = 1.0$

$I_d = 0.26 \ldots 0.29$

Figure 3.16: Incremental displacements for a piston advance from 1.25 to 1.00 mm for different C/D and I_d values with material $S1$

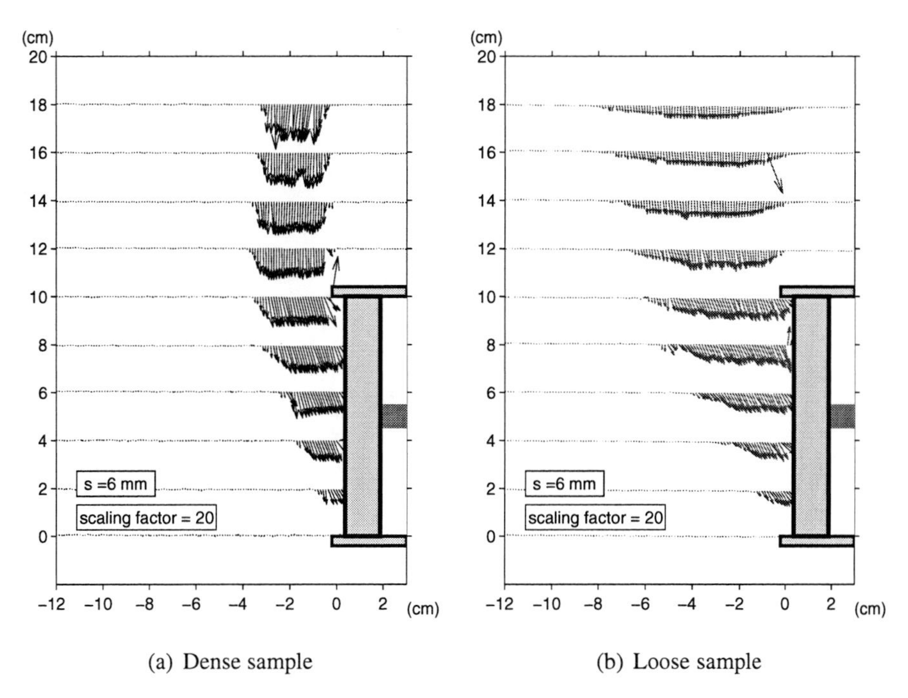

(a) Dense sample

(b) Loose sample

Figure 3.17: Distribution of incremental displacement vectors in horizontal cuts through the specimen

in "noise" of the derived quantities. Therefore, the author concentrated on qualitative interpretations of the results.

Incremental shear strains

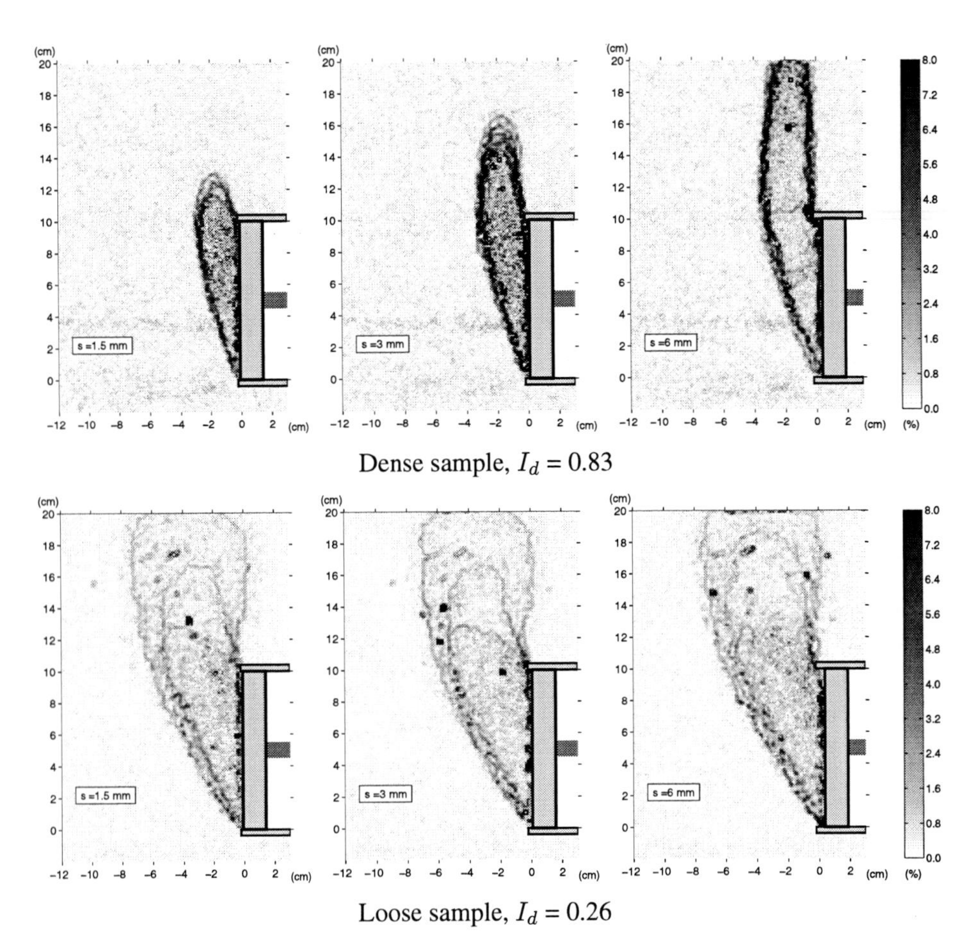

Figure 3.18: Shear strains I_2^ε for different advance steps and different relative densities in material $S1$

Fig. 3.18 shows the shear strains I_2^ε for dense and loose samples at various steps of piston advance.[9] The dense samples showed a well-defined concentration of shear strains (dark lines). The concentration zones form an arch ahead of the tunnel. Single features extend like "branches" above the arch.

For the loose samples shear strains were distributed in rather large shear *zones*, which also show some minor shear strain concentrations, though.

[9]The single spots of "shear strain", especially in the pictures for the loose samples, are numerical artefacts due to outliers.

Curl of the displacement field

The curl of the displacement field is a measure of (microscopic) circulation around the out-of-plane axis in every point. It should not be confused with the rotation of particles, which cannot be evaluated with the applied PIV technique. Still, an evaluation of (3.11) helps to visualise the soil movements (Fig. 3.19).

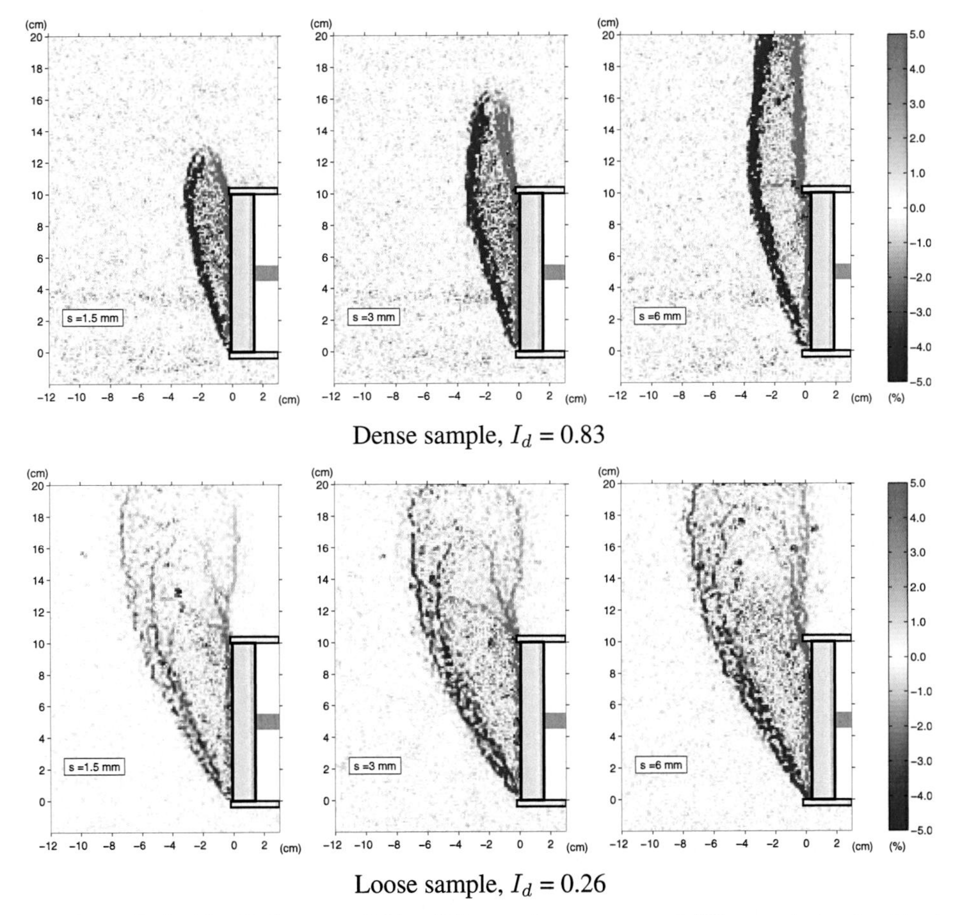

Loose sample, $I_d = 0.26$

Figure 3.19: Curl of the displacement field at different advance steps and different relative densities in material $S1$

The red colour in Fig. 3.19 indicates a counter-clockwise, the blue colour a clockwise circulation of the vector field. For the dense samples a concentration of circulation in well-defined zones can be observed. A circulation band indicating clockwise circulation emanates from the bottom of the tunnel and heads towards the ground surface. A corresponding band with counter-clockwise circulation heads from the tunnel crown upwards. The two bands touch above an arch that is formed ahead of the tunnel.

The loose samples show a tree-like pattern of circulation that extends into the soil domain in motion and does not change with increasing piston displacement.

Evolution of void ratio

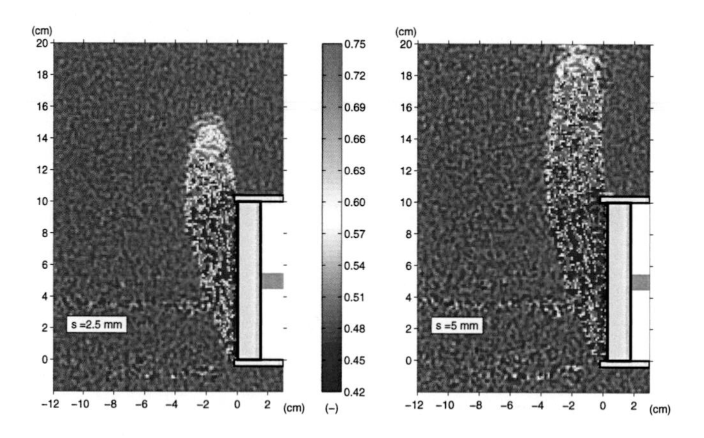

(a) Dense sample, $e_0 = 0.48$

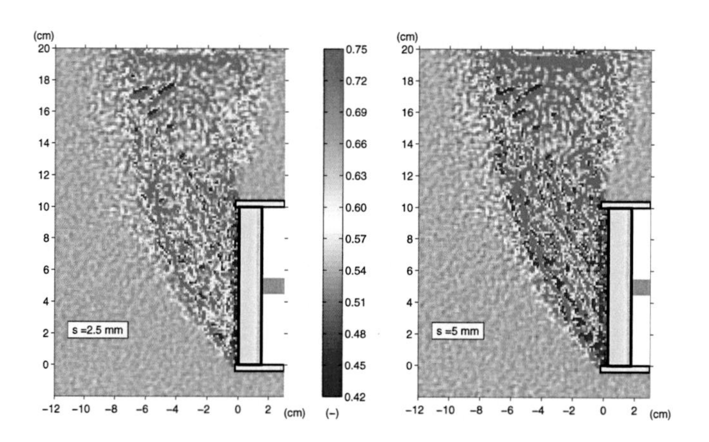

(b) Loose sample, $e_0 = 0.66$

Figure 3.20: Plot of void ratio e at different advance steps

Fig. 3.20 a shows a plot of void ratio for different advance steps of a dense soil. It becomes clear, how the initially dense sample changed its density ahead of the face and how it loosened in a well-defined chimney towards the soil surface.

In an initially loose soil (Fig. 3.20 b), the soil loosened further in parts of the soil domain, in others it densified. The scatter of void ratios between e_{min} (dark blue) and e_{max} (dark red) does not seem to have physical reasons. Probably the integration of a strain measure over a number of evaluation steps enhances the noise of the evaluation too much.

Inclination of the wedge

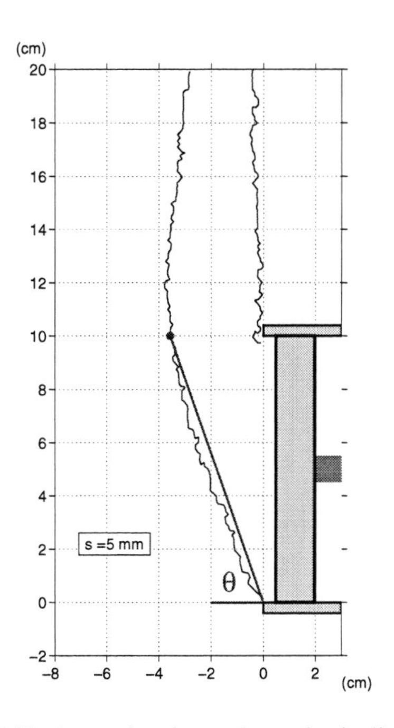

Figure 3.21: Approximation to the wedge inclination θ

The inclination of the moving soil wedge in front of the tunnel face was also evaluated. For each picture of incremental displacements the contour between soil at rest and soil in motion was determined with MATLAB (contour value: $|u| = 0.05$ mm). The intersection of this contour with a horizontal line at the tunnel crown level (y = 10 cm) was connected to the origin (straight line in Fig. 3.21). The angle θ between this line and the horizontal served as an approximation to the inclination of the soil wedge. Strictly speaking, the contour is slightly curved and not a straight line. For a qualitative statement on the wedge inclination the linearisation seemed acceptable, though.

Fig. 3.22 a shows the obtained wedge inclinations for dense and loose samples vs. a piston displacement from 1.25 to 9.0 mm for material $S1$. For the dense samples θ was approximately 73° for a piston displacement of 1.25 mm. With more piston displacement the wedge inclination decreased to 68°. The loose samples, in contrast, showed a much lower wedge inclination around 60°, which did not show a clear dependency on piston displacement.

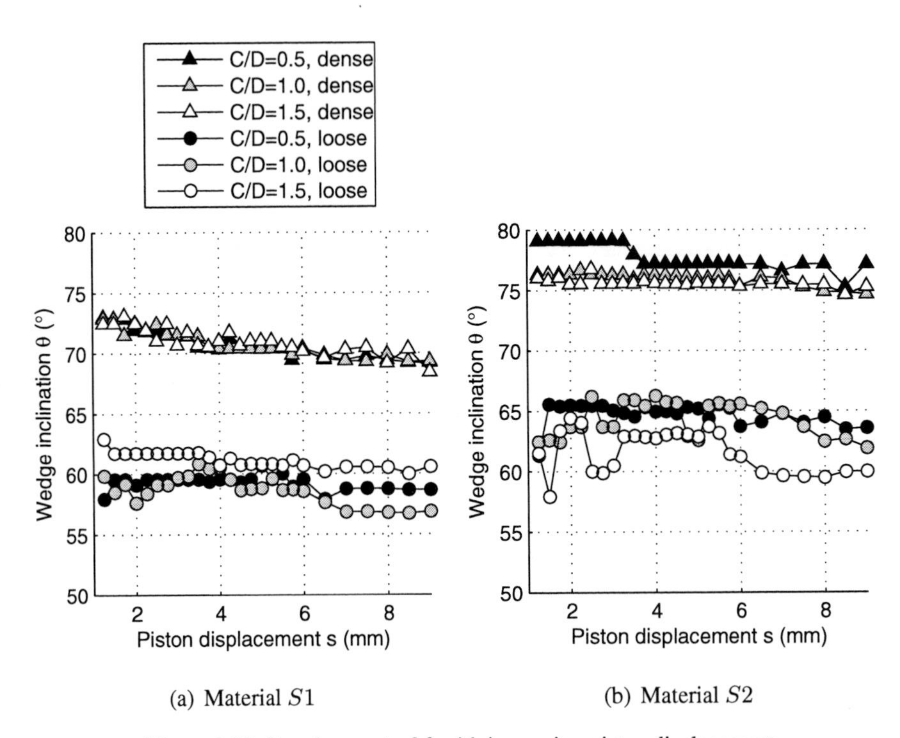

(a) Material $S1$ (b) Material $S2$

Figure 3.22: Development of θ with increasing piston displacement

3.2.5 Interpretation of results

Dense sand

The shape and extent of the observed failure zone is in good qualitative agreement to centrifuge experiments by *Chambon* and *Corté* (Fig. 2.13), *Kamata* and *Mashimo* (Fig. 2.18), and 1g model tests by *Takano* et al. (Fig. 2.22).

The PIV analyses show that the overburden had a negligible influence on the shape and extent of the failure zone for dense soil samples. The impact of density was much more pronounced: in dense sand the failure zone developed stepwise towards the ground surface. As soon as it reached the ground surface, a chimney-like and a wedge-like part could be distinguished (Fig. 3.23). This configuration resembles the analytical *Horn* model (cf. Sec. 2.1.1).

There is one important restriction, though: no intermediate shear bands were detected at the crown level of the model tunnel. The observations rather indicate a zone between chimney and wedge, in which the incremental displacement vectors continuously changed their orientation.

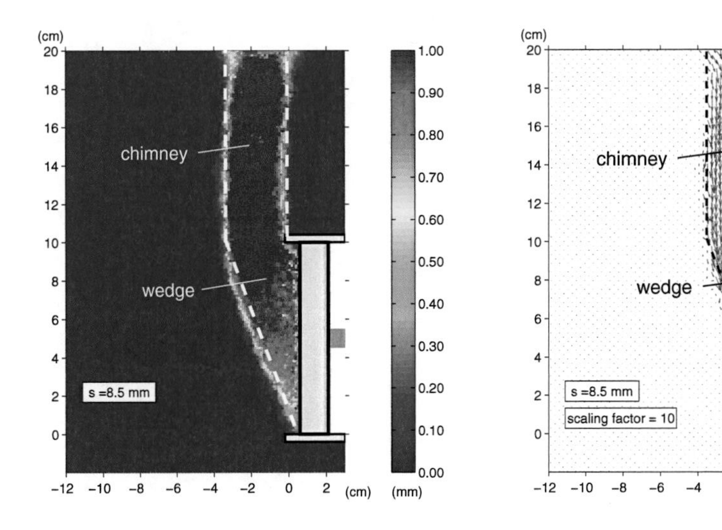

(a) Colour plot of incremental displacements (b) Vector plot of incremental displacements

Figure 3.23: *Horn* analogy of the failure zone for a piston advance step from 8.0 to 8.5 mm (C/D = 1.0; I_d = 0.9; $S1$)

Therefore, *Horn*'s assumption of a *rigid-block* mechanism is not supported by the author's experiments. But it is an open question whether the energy dissipated in the observed shear zone is more or less equal to that dissipated in a discrete shear plane.

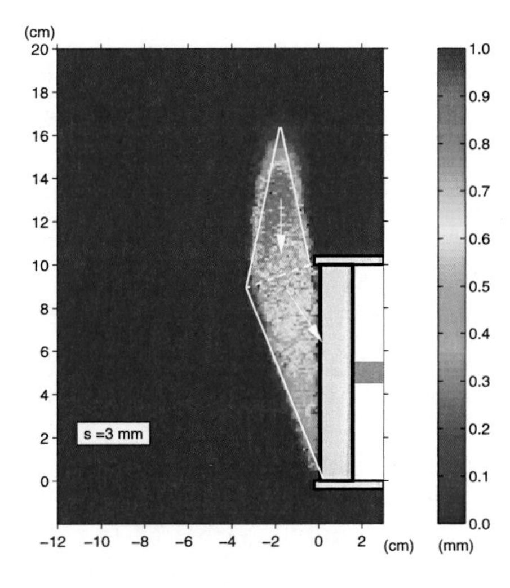

Figure 3.24: Shape of the failure mechanism predicted by the *Léca* and *Dormieux* model

The **trapdoor problem** serves to study the load transfer from a yielding structure (trapdoor) to an adjacent non-yielding support. Usually the relation between trapdoor support force P and displacement s is investigated. The adjacent sketch illustrates the difference between the active case (i.e. lowering the trapdoor) and the passive case (i.e. the trapdoor is pushed into the soil).
There is a large amount of literature on the topic. The interested reader is referred to *Dewoolkar* et al. [32], *Ono* and *Yamada* [97] or *Graf* [47].

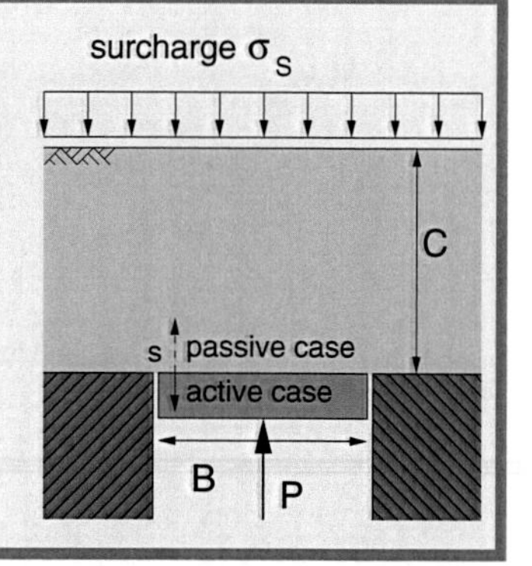

The *inclination* of the observed wedge was also compared with predictions of the *Horn* model. For this purpose, a number of *Horn* configurations were investigated with $\varphi_c \approx 32°$ (for the loose samples) and $\varphi_p \approx 42°$ (for the dense samples).

The theoretical predictions for θ_c vary between 62° and 76° for calculations with φ_c; with φ_p a range θ_p between 68° and 83° is obtained. The lower predictions fit the experimental results quite well (Fig. 3.22 b).

The inclination $\vartheta = 45° + \varphi/2$, which is adopted in some publications (e.g. *Katzenbach* and *Strüber* [65]) is a good approximation for the wedge inclination of the loose samples.

Taking the three-dimensional nature of the failure mechanism into account, also the mechanism by *Léca* and *Dormieux* (Fig. 3.24) seems quite realistic.

Arching and trapdoor analogy

When looking at the shape of the failure zone and its development, the term *arching* comes to mind. In a general sense, arching means the redistribution of stresses around an underground structure. Arching has been studied experimentally, theoretically and numerically by means of the trapdoor problem. It has been found that, even for small displacements, the soil load on the trapdoor becomes substantially different from the initial geostatic loads. The degree of arching depends on various factors, such as stiffness of soil and structure, geometry, overburden height and strength of the soil, to name just a few. In the following, some findings for the active trapdoor problem will be put into context with the PIV evaluations of the author's experiments.

Graf [47] investigated the active trapdoor problem experimentally. By means of X-rays he identified zones of density inhomogeneities during lowering of the trapdoor. His experimental setup and the X-ray observations for s = 6 mm are sketched in Fig. 3.25. Shear bands (dark lines) form an arch above the trapdoor, similar to those observed in the PIV results (Fig. 3.25 b).

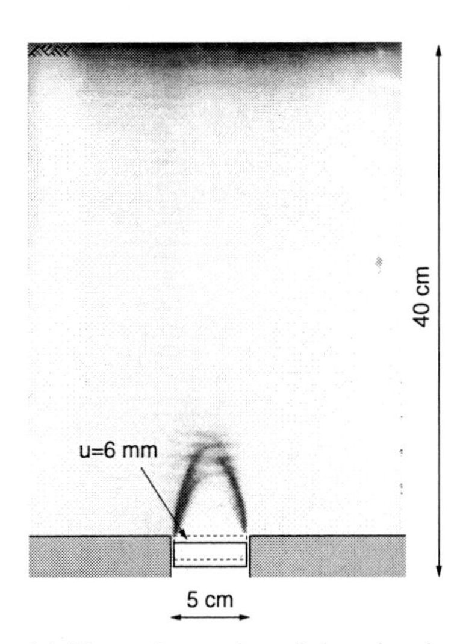

(a) X-ray observation of shear bands in dense sand, adapted from [47]

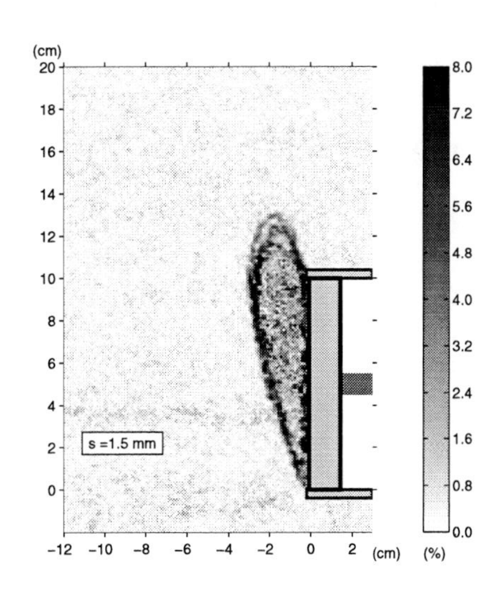

(b) Colour plot of shear strains, dense sample, $C/D = 1.0$

Figure 3.25: Trapdoor analogy

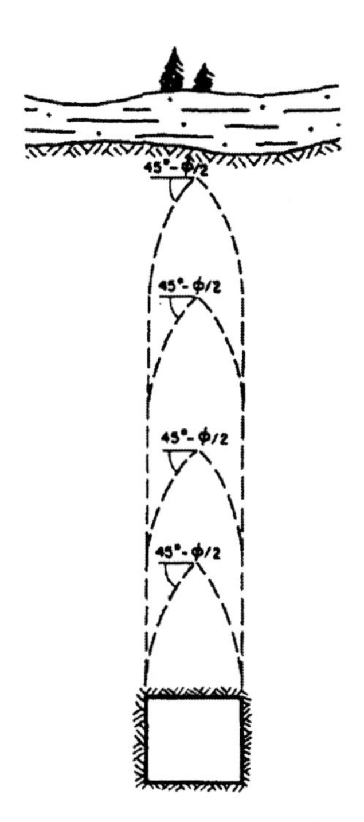

(a) Sketch of progressive chimney cave development, [15]

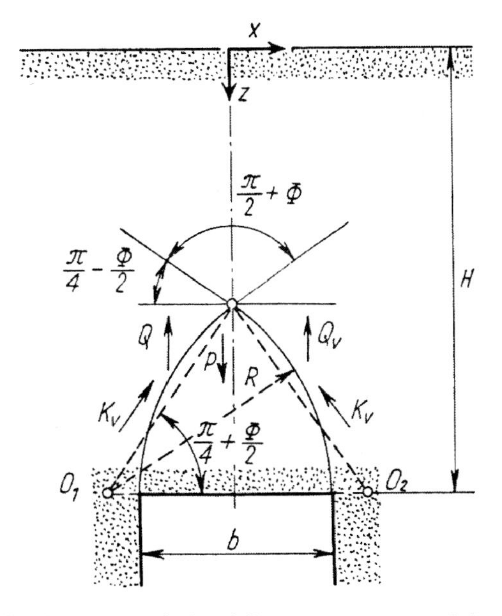

(b) Geometry of the failure zone suggested by *Balla*, [11]

Figure 3.26: Arching models for plane strain

For plane strain conditions, some arching theories have been put forward: Fig. 3.26 a shows a sketch of a *chimney caving*, a term from mining engineering. It starts with instabilities at the tunnel crown and proceeds towards the surface, if a stable self-supporting arch cannot be formed. An identical shape of the failure zone (Fig. 3.26 b) was proposed by *Balla* [11]. He, arbitrarily, chose circular arcs as boundaries of the failure zone. In both cases the slip lines have vertical tangents at the tunnel crown level, and intersect each other at an angle of $\pi/2 - \varphi$.

The behaviour of a granular soil above a trapdoor resembles very much what was observed in the PIV evaluations for the dense samples. To the author's knowledge there are only few publications on the active trapdoor problem which link the propagation of the failure zone with the support force underneath the trapdoor, e.g. *Dewoolkar* et al. [32].

Propagation of the failure zone

As a results of the experimental observations a simple model for the kinematics in the dense soil was developed. It explains how "fast" the failure zone propagates towards the soil surface.[10]

In plots of void ratio e (Fig. 3.27 a), a transition zone between dense and loose soil could be observed, forming an arch that propagates from the tunnel crown level to the soil surface (Fig. 3.27 b).

Assume this transition zone to be a thin horizontal discontinuity between sand grains in motion ($v_2 = V$) and those unmoved ($v_1 = 0$). The discontinuity moves with a velocity u vertically downwards (Fig. 3.27 c).

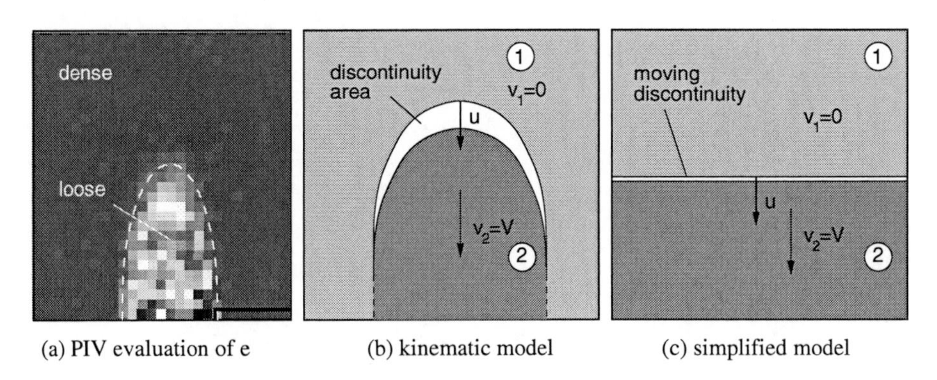

(a) PIV evaluation of e (b) kinematic model (c) simplified model

Figure 3.27: Kinematic model for the propagation of the failure mechanism

[10] A very similar approach was applied by *Drescher* and *Michalowski* [37] for the description of pseudo-steady flow through a plane hopper.

The mass balance equation can be written as

$$[\rho_i(v^i - u)] = 0 \quad . \tag{3.17}$$

Expressing the density ρ in terms of porosity n

$$\rho_i = (1 - n_i)\, \rho_s \quad , \tag{3.18}$$

and dividing by the constant unit weight of the grains ρ_s, (3.17) can be rewritten as:

$$\begin{aligned} (1-n_1)(0-u) &= (1-n_2)(V-u) \quad , \\ \rightsquigarrow u &= \frac{1-n_2}{n_1-n_2} V \quad . \end{aligned} \tag{3.19}$$

Note that u and V have opposite signs, if $n_1 < n_2$.

In terms of void ratio e, the propagation velocity u can be expressed as

$$u = \frac{1+e_1}{e_1-e_2} V \quad . \tag{3.20}$$

To check this model's performance, different advance steps for a test with dense sand and a C/D = 1.0 were evaluated with respect to void ratio e and incremental displacements $\mathbf{u}$. Fig. 3.28 shows contours of the failure zone at piston advance steps between s = 0.5 mm and s = 4.75 mm. These contours mark the border between the soil at rest and the soil in motion (threshold value u = 0.05 mm).

For the given example, the moving discontinuity propagates vertically *upwards* with a "velocity" $u \approx$ -5.3 mm per increment. From the vector plots of incremental displacements, the "velocity" V of the grains in motion can be approximated by $V \approx$ 0.72 mm per increment.

The mean void ratio in the soil at rest (domain 1 in Fig. 3.27) is e_1 = 0.48. In the chimney it is not so easy to determine e_2, because the values show a reasonable amount of scatter. Fig. 3.29 illustrates how e_2 was quantified: at different advance steps in which the chimney had already developed, e was evaluated for each cell in a certain interrogation window. All values for e in this domain were then averaged to a mean value $\bar{e}$ for this advance step. An evaluation of several steps lead to an approximate value for $e_2 \approx$ 0.66.

With the above values, (3.20) reads:

$$\begin{aligned} u &= \frac{1+e_1}{e_1-e_2} V \quad , \\ \rightsquigarrow u &= \frac{1+0.48}{0.48-0.66} \cdot 0.72 \text{ mm} \\ &= -5.9 \text{ mm} \quad . \end{aligned}$$

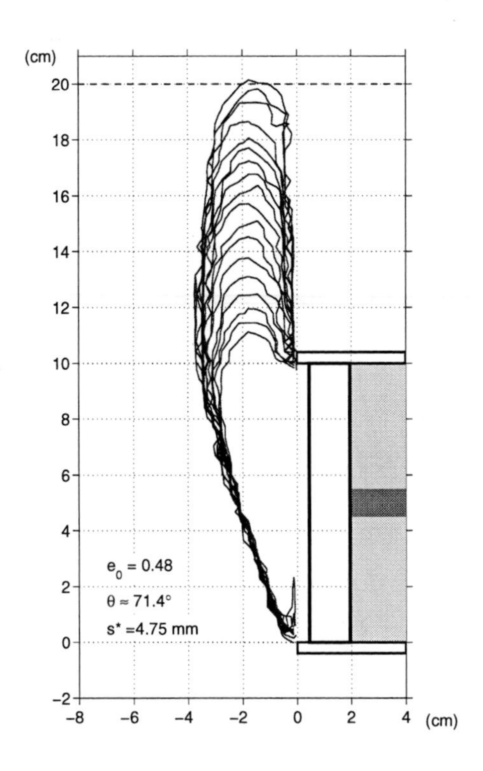

Figure 3.28: Contours of the loosening zone for different advance steps (C/D = 1.0, material $S1$)

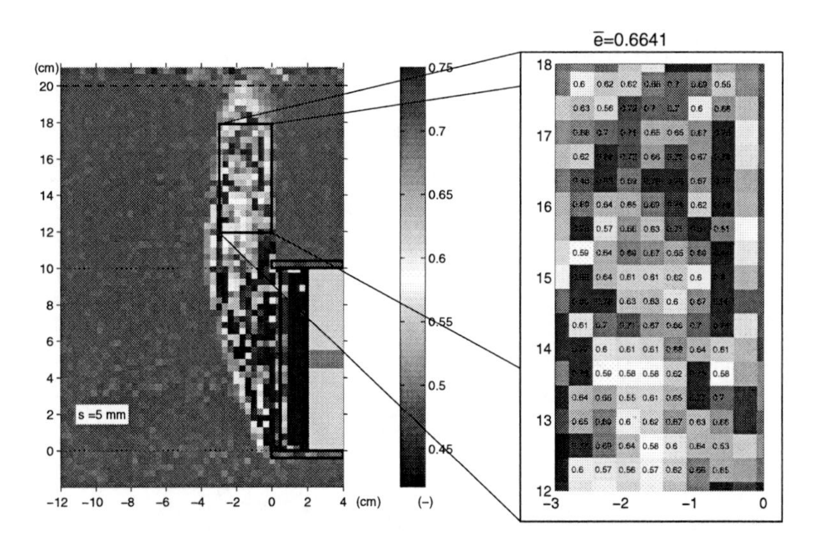

Figure 3.29: Evaluation of void ratio e in the silo body

This result is a fairly good approximation to the measured one (-5.3 mm). Keeping in mind that only average values for e_2 and V can be determined, (3.20) provides a reasonable estimate for the propagation velocity of the failure zone. But due to the scaling effect, a transfer to the prototype scale must be treated with caution (cf. *Technical Committee TC2* of ISSMGE [124]).

Loose sand

For the loose samples, distinct shear planes were not detected in the PIV evaluations. More precisely, a rather diffuse failure zone was observed, which did not significantly change its shape throughout the test. These statements are in agreement with findings by *Vardoulakis* et al. [128], who did not find shearbands in active trapdoor problems with loose sand.

The shape of the failure zone does not resemble any of the proposed models, described in Chap. 2.

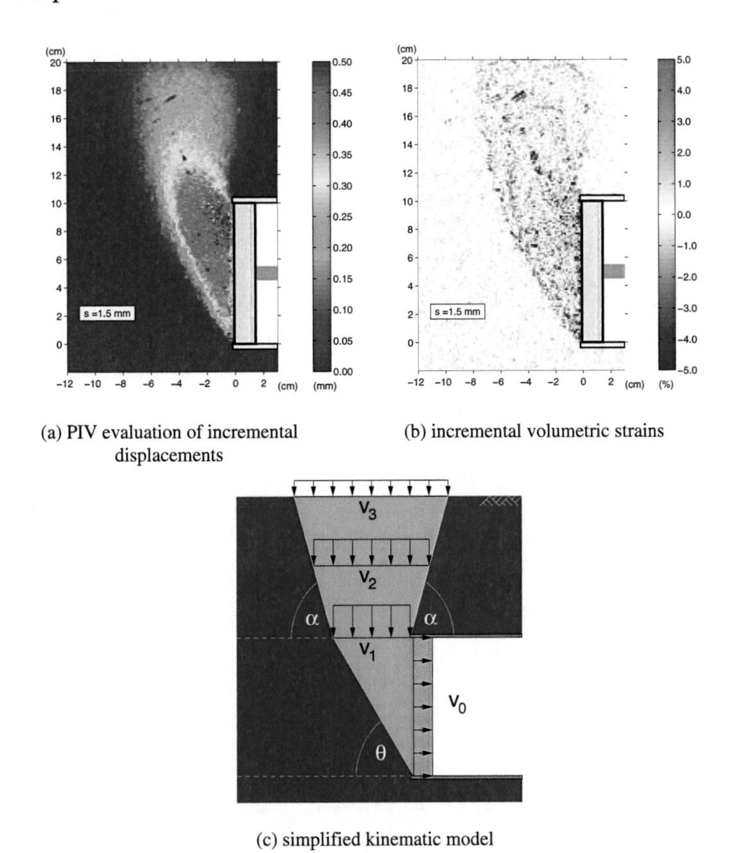

(a) PIV evaluation of incremental displacements

(b) incremental volumetric strains

(c) simplified kinematic model

Figure 3.30: Kinematics of failure zone for loose sand

Kinematics of the failure mechanism

A possible reason is that a loose grain structure has less ability to sustain arches. Fig. 3.30 a indicates that there is only little arching in the soil (orange zone close to the face). With increasing distance from the tunnel crown the magnitude of the incremental displacements decreases, but the width of the trough increases. In other words, more particles are involved in the movement by comparison with the dense samples. The derived incremental volumetric strains (Fig. 3.30 b) show that the zone of soil in motion does not change its volume significantly.

The observed behaviour resembles that of an incompressible fluid in a funnel: for continuity reasons the product of average velocity and funnel diameter remains constant, i.e. $v_0\ A_0 = v_i\ A_i$. Of course, soil is not incompressible. But a loose sample with a void ratio close to the critical one shows little tendency to change its density when sheared.

Scaling effect

According to *Stone* and *Muir Wood*, “in order to maintain geometric similarity between the kinematic deformation mechanisms observed in models using sands with different particle sizes, the physical dimensions and boundary movements of the models should also be scaled in proportion to the particle size [117]”. In the author’s experiments the dimensions of the model were not scaled, but it is possible to normalise the piston displacement, s, with the mean diameter of the sand particles d_{50}. *Stone* and *Muir Wood* mentioned that similar stages in the displacement pattern could be observed for s/d_{50} = const. Therefore, PIV results for materials $S1$ and $S2$ for $s/d_{50} \approx 6.0$ are presented in Figs. 3.31 and 3.32.

The extents of the failure zones towards the soil surface, for a given s/d_{50}, were in fair agreement for both materials at different C/D ratios. For the fine sand, the width of the chimney was considerably smaller, though. Also the inclination θ of the wedge was steeper (cf. Fig. 3.22 b) with an average of 76° for the dense samples and 64° for the loose samples. Moreover, the absolute values of incremental displacements were higher for material $S2$ (indicated by the dark red colours in Fig. 3.32).

Interpretation of differences

The failure mechanism for material $S2$ showed a noticeably larger inclination of the wedge, even though the respective friction angles are roughly equal. The reason for this different behaviour must be attributed to the scaling effect. Therefore, *quantitative* conclusions from model behaviour to prototype behaviour are hardly possible *for the dense samples*.

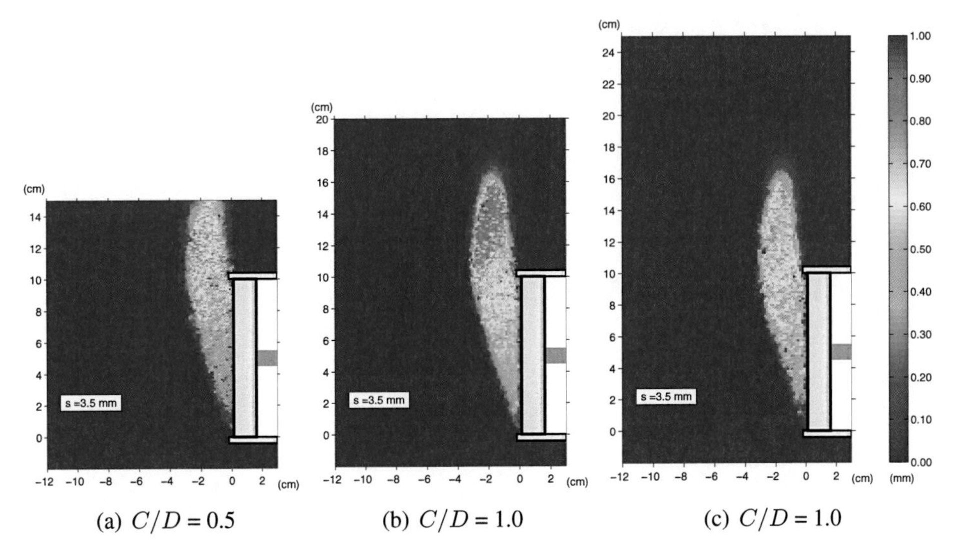

(a) $C/D = 0.5$ (b) $C/D = 1.0$ (c) $C/D = 1.0$

Figure 3.31: Incremental displacements for material $S1$, dense samples with I_d = 0.80 ... 0.85, s/d_{50} = 3.5 mm/0.58 mm $\approx$ 6.0

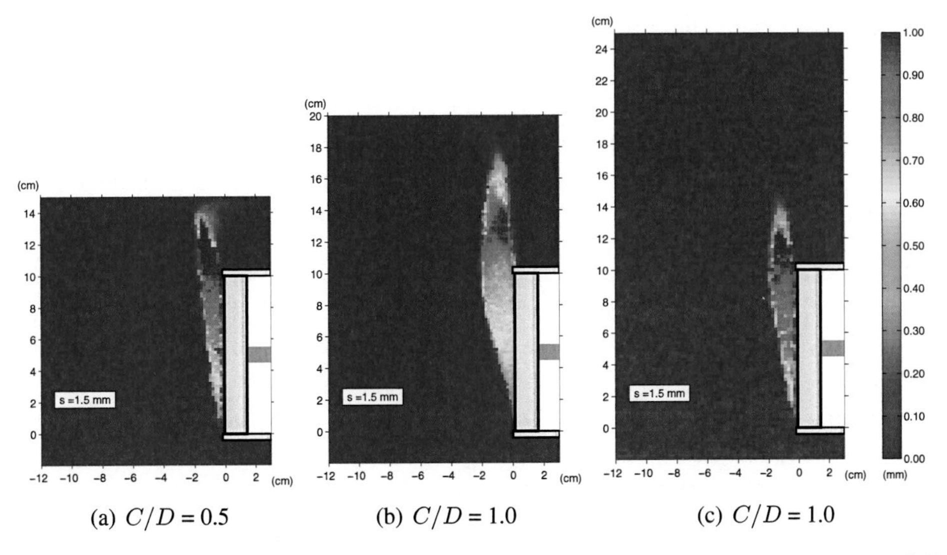

(a) $C/D = 0.5$ (b) $C/D = 1.0$ (c) $C/D = 1.0$

Figure 3.32: Incremental displacements for material $S2$, dense samples with I_d = 0.86 ... 0.98, s/d_{50} = 1.0 mm/0.24 mm $\approx$ 6.0

But it is important to mention that the *qualitative* behaviour and general shape of the failure zone did not depend on grain size. Also, above restriction does not hold for the loose samples.

3.3 Quantitative investigation of the necessary support force

To validate the proposed models for the necessary support force/pressure, the model box was modified to allow for measurements of the resulting axial force on the piston. One purpose was to relate the development of the support pressure to the PIV evaluations.

Again, the cover-to-diameter ratio C/D and the density I_d of the samples were varied. The influence of grain size was investigated by using materials $S1$ and $S2$.

3.3.1 Experimental setup and tested materials

Sandbox and tunnel model

The model box for the second series of experiments is schematically sketched in Fig. 3.33, a picture of the box is displayed in Fig. 3.34. The tunnel was modelled with a hollow aluminium cylinder with an inner diameter of 10 cm and a wall thickness of 4 mm. As for the PIV measurements, the model tunnel protruded approx. 7 cm into the soil domain (Fig. 3.33 a). The distance between cylinder wall and bottom of the box was 5 cm. Thus, an influence of the bottom boundary on the force measurements could be ruled out. As a result of the PIV investigations, the dimensions of the box were considered large enough.

Wall friction was not considered to play a significant role and, therefore, no explicit investigation of wall friction on the side panels of the box was executed to evaluate its influence on the force measurements. This assumption is supported by *Hauser* [50], who published force measurements from small-scale model tests for coffer dams. With a comparable geometry, different configurations of wall friction (with and without teflon sheets between sand and side walls) did not show a significant influence on the measured force.

The face of the model tunnel was supported by an aluminium disk with a slightly smaller diameter than the inner diameter of the tunnel (D_{disk} = 9.8 cm), thus eliminating friction between disk and tunnel. The piston rod was supported by a linear roller bearing, embedded in the side wall. It allowed for a horizontal movement of the piston into (and out of) the model tunnel.

The rod made contact with a miniature load cell[11] that was mounted on a sliding carriage on the outside of the side wall. Because of the low stresses expected in the model the load cell had a nominal load of only F_{nom} = 10 N.

[11] Trade name: Miniatur-Druckkraftsensor, 'burster, type 8413'

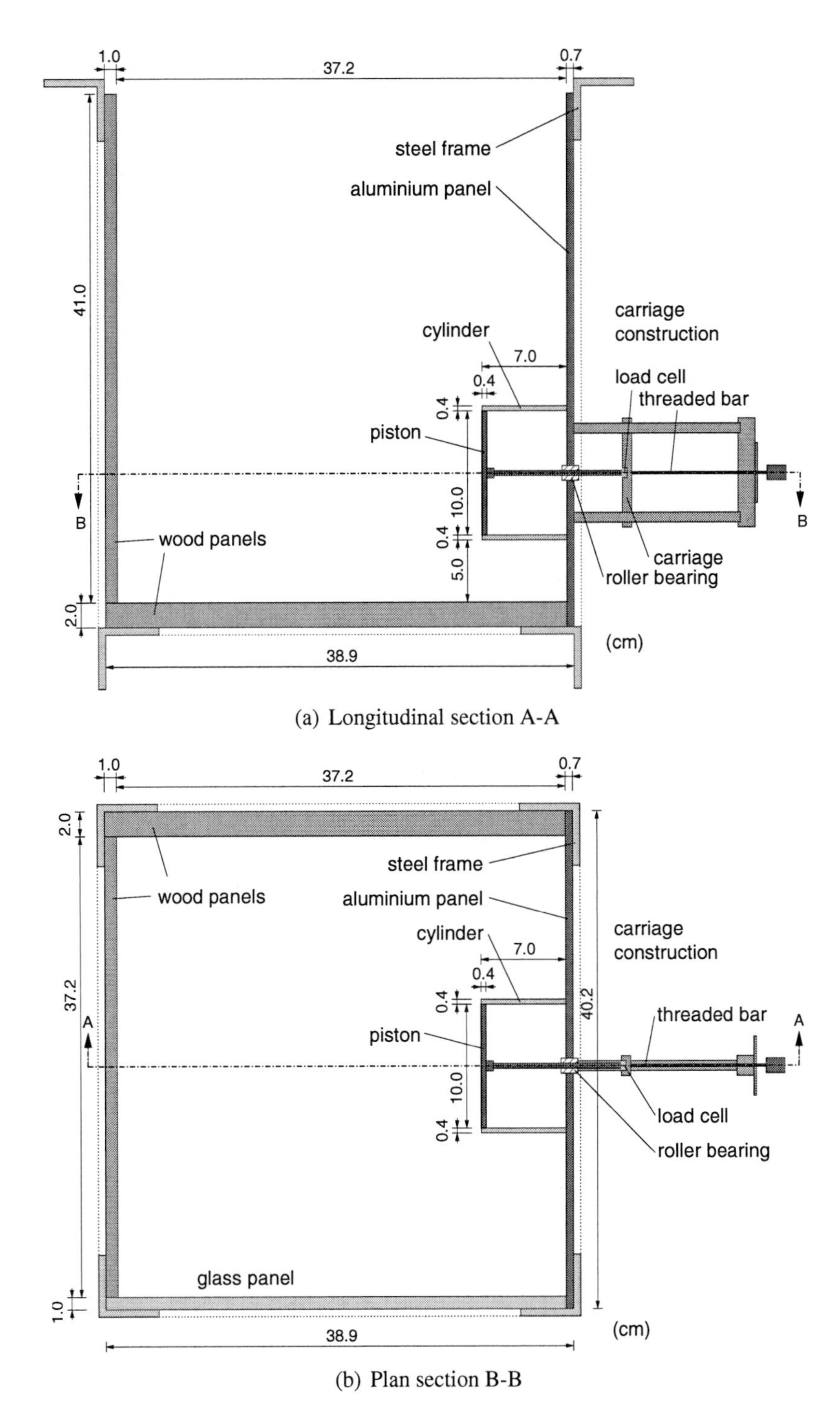

(a) Longitudinal section A-A

(b) Plan section B-B

Figure 3.33: Experimental setup for the second set of experiments

Figure 3.34: Box and model tunnel for the second set of experiments

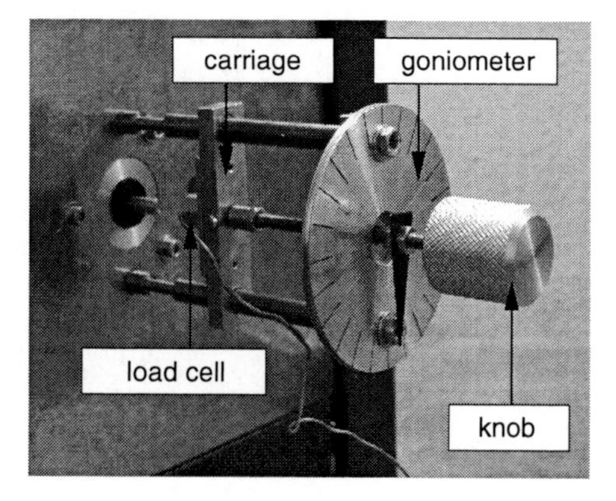

Figure 3.35: Carriage construction with load cell, goniometer and turning knob

The carriage could be moved by turning a knob (Fig. 3.35). A full revolution of the threaded bar led to a carriage displacement of 1 mm. In order to control very small incremental displacements, a goniometer[12] was used.

Consideration of friction in the system

It was crucial to prevent any sand ingress into the gap between piston and model tunnel. This would have led to an immediate obstruction of the piston, and thus stopped the experiment. In preliminary sandbox tests different methods were investigated to minimise friction in the piston-tunnel-system (*Doro* [36]):

- a thick film of bearing grease in the gap (Fig. 3.36 a),

[12]A thin aluminium disk with a graduation of 15° and an indicator were attached to the carriage construction.

- bearing grease in the gap in combination with a temporary cling foil cover over the gap (Fig. 3.36 b),
- complete cover of the outside of the model tunnel with cling foil (Fig. 3.36 c).

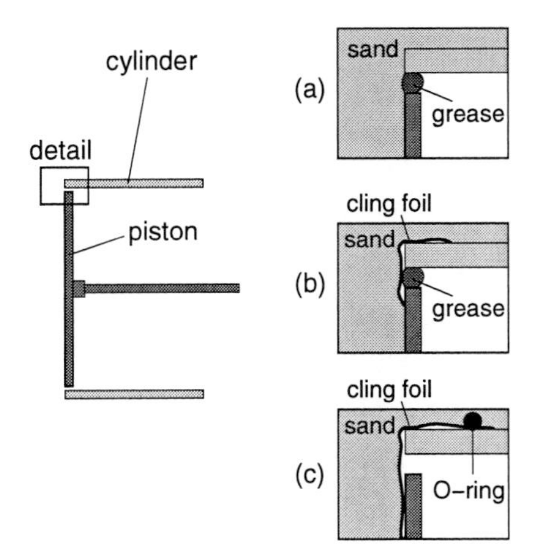

Figure 3.36: Investigated sealing methods for the gap between piston and model tunnel

Figure 3.37: Model tunnel with cling foil protection against sand ingress

The first two methods could not be applied successfully, because the registered force dropped to zero after the first couple of advance steps. This was probably due to sand grain penetration into the gap and low viscosity of the grease. The best results were obtained with a complete cover of piston and cylinder with cling foil and application of talcum powder between foil and piston (Fig. 3.37).

Fig. 3.37 illustrates how the cling foil was applied to the cylinder: it was fixed with a rubber O-ring, after a certain clearance in the order of 1 to 2 cm was provided for. In this way there was some space for the sand to move into the cylinder before the earth pressure was taken over by the cling foil. In the author's opinion, above procedure allowed for a sufficient range of displacements for the purpose of the investigation.

Reference measurements with static water table

To quantify the friction within the whole system, reference measurements were made with hydrostatic water pressure (cf. *Ravazzini* [105]). The idea of the reference measurements was to apply a *follower load*, i.e. a load that remained constant while moving the piston. Thus, a difference between applied and measured force could be interpreted as overall friction.

The wall with the model tunnel was turned inside out, and a water supply system was attached to the aluminium cylinder (Figs. 3.38 and 3.39). A perspex cylinder was used to avoid contact between rubber membrane and piston.

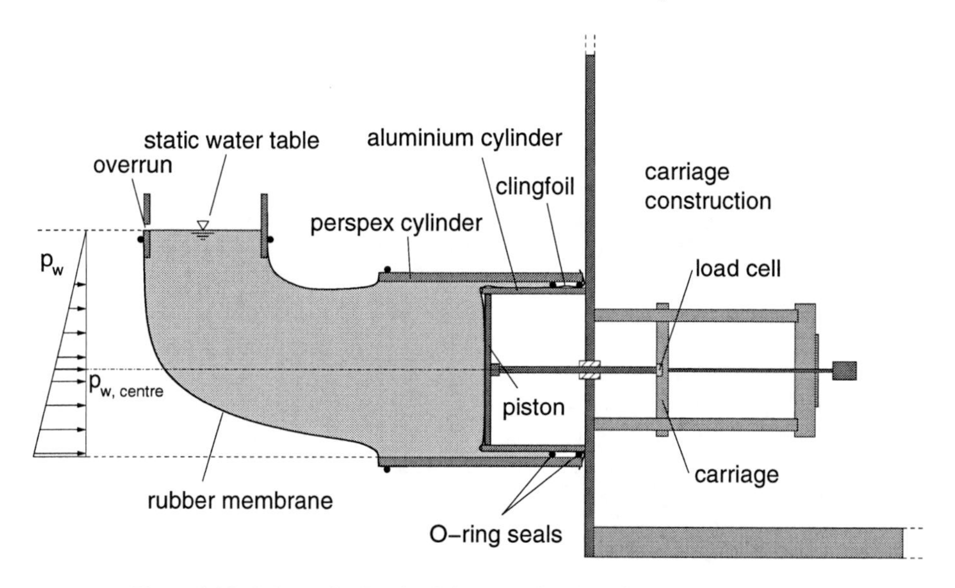

Figure 3.38: Schematic sketch of the setup for the reference measurements

The gap between piston and tunnel wall was sealed by a layer of cling foil and two O-rings. Thus, the same sealing conditions applied to, both, sand and water experiments.

Inside the membrane-cylinder system, a constant water pressure was maintained by means of an overrun at a predefined level. The force acting on the piston was calculated to be:

$$F_{\text{nom}} = p_{w,\text{centre}} \cdot A_{\text{piston}} \quad , \tag{3.21}$$

with the water pressure $p_{w,\text{centre}}$ at the centre of the piston.

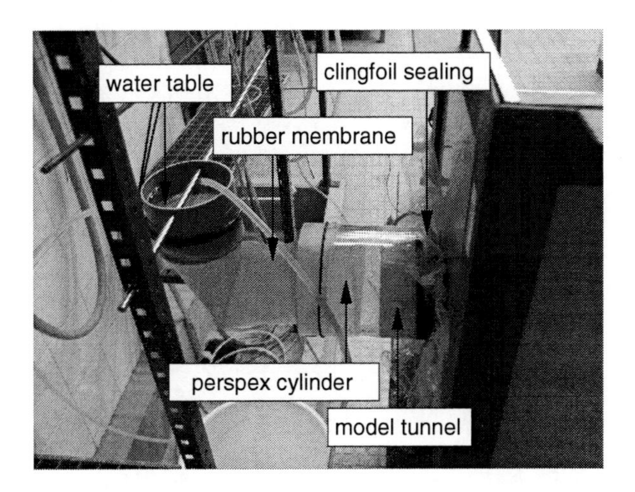

Figure 3.39: Setup for the reference measurements

Two reference measurements with slightly different water tables at tunnel crown level were made (Fig. 3.40). It became obvious that the force, registered by the load cell was lower than F_{nom}. The difference between F_{nom} and F_{measured} was attributed to friction in the roller bearing and influence of the cling foil.

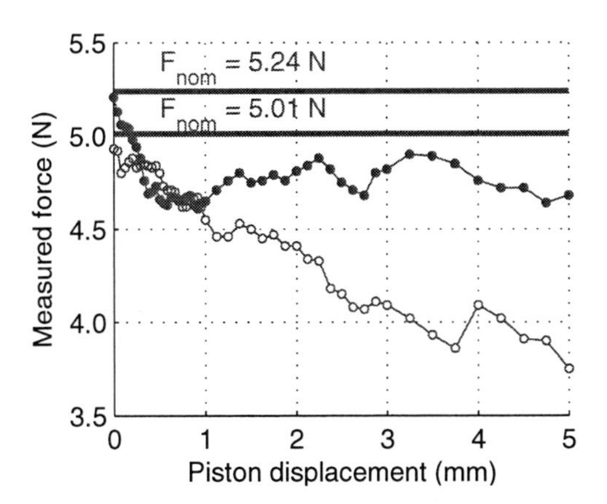

Figure 3.40: Influence of friction in the system

As a result corrective terms ΔF_1 and ΔF_2 were derived, ΔF_2 as a function of piston advance s (cf. Fig. 3.41):

$$\Delta F_1 \text{ (N)} = 0.5 \quad , \tag{3.22}$$

$$\Delta F_2 \text{ (N)} = 0.2 + 0.2\, s \text{ (mm)} \quad . \tag{3.23}$$

All measurements of the second test series, which are shown in the following, were corrected with ΔF_1 (full circles) *and* ΔF_2 (empty circles). The "true" load-dis-

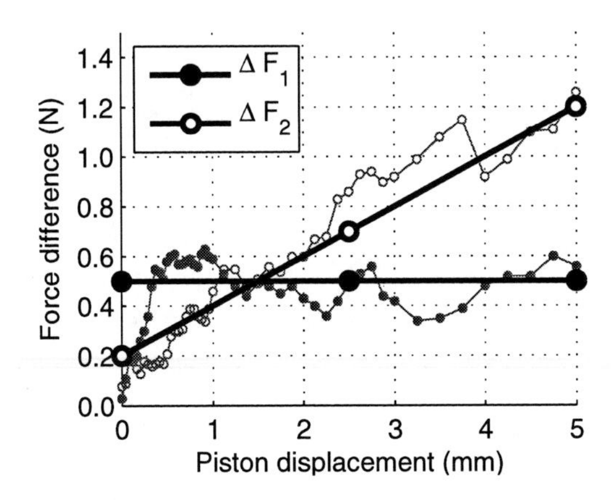

Figure 3.41: Proposed correction functions

placement curves were expected in the range between $F_1 = F_{\text{measured}} + \Delta F_1$ and $F_2 = F_{\text{measured}} + \Delta F_2$.

3.3.2 Test procedure

Preparation of the sand body

In order to compare PIV evaluation and force measurements, the soil was prepared in the same way as described in Sec. 3.2.1.

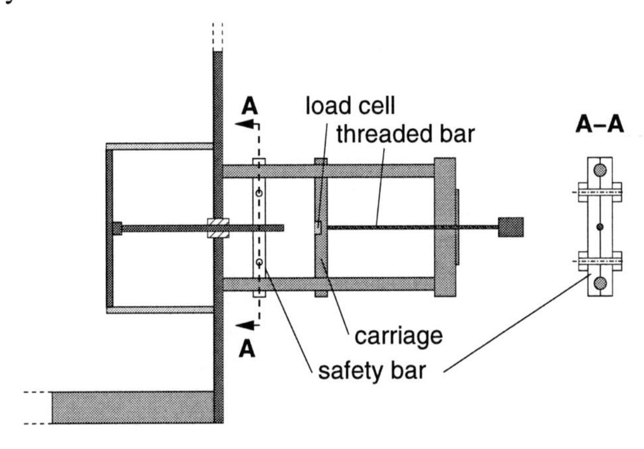

Figure 3.42: Illustration of safety bar application during sample preparation

The load cell was very sensitive to overloading. Therefore, a safety bar (Fig. 3.42) was applied to the carriage construction, which temporarily inhibited any contact between piston rod and load cell during sample preparation. Before each test the carriage with the load cell was slowly moved into contact with the rod and, finally, the safety bar removed.

Throughout the test series the load cell was connected to an amplifier which displayed the current force on the cell. Before every test series the load measuring system was calibrated for a range between 0 and 8 N, by turning the tunnel in a vertical direction and putting dead loads on the piston.

Steps of the model tests

The experiments were, again, performed displacement-controlled, by incrementally retracting the carriage. The piston rod was in contact with the carriage, and thus the load cell, which measured the resulting force exerted by the ground on the piston.

For the first millimetre of advance, displacement increments of $\Delta s = 0.042$ mm were applied. Thus, the force reduction for the very first displacements could be captured well. For $s > 1$ mm, the incremental advance was set to $\Delta s = 0.125$ mm.

Test programme

Test No.	C/D	material	density	I_d
1-2	0.25	$S1$	dense	0.79, 0.81
3-4	0.25	$S1$	loose	0.22, 0.35
5-6	0.25	$S2$	dense	0.73, 0.76
7-8	0.25	$S2$	loose	0.07, 0.09
9-12	0.50	$S1$	dense	0.70, 0.74, 0.84
13-17	0.50	$S1$	loose	0.25, 0.34, 0.35
18-21	0.50	$S2$	dense	0.69, 0.71, 0.74
22-26	0.50	$S2$	loose	-0.01, 0.01, 0.10
27-28	0.75	$S1$	dense	0.79, 0.83
29-30	0.75	$S1$	loose	0.29
31-32	0.75	$S2$	dense	0.69, 0.82
33-34	0.75	$S2$	loose	-0.03, 0.11
35-38	1.00	$S1$	dense	0.70, 0.73, 0.82
39-40	1.00	$S1$	loose	0.33
41-42	1.00	$S2$	dense	0.73
43-44	1.00	$S2$	loose	0.08, 0.09
45-46	1.00	$S1$	dense	0.77, 0.80
47-48	1.00	$S1$	loose	0.25, 0.29
49-50	2.00	$S1$	dense	0.76
51-52	2.00	$S1$	loose	0.22, 0.29

Table 3.4: Test programme for the force measurements

For each combination of C/D and density I_d at least two separate tests were performed to check the reproducibility of the results. An overall number of 52 tests were performed (Tab. 3.4) with C/D = 0.25 ... 2.0 and (initially) dense and loose samples.

3.3.3 Results

Review: Scaling laws

As forces were measured in the second test series, the relations between force measurements on different scales had to be taken into consideration.

Görtler [46] stressed the point that the first essential step of a thorough dimensional analysis is the identification of governing quantities for a given problem. The following quantities were initially assumed to govern the magnitude of the necessary support force:

- geometry:
  - diameter D of the model tunnel ,
  - soil cover C ,
  - piston advance s ;
- soil properties:
  - dry self-weight of the soil γ_d ,
  - density index I_d ,
  - critical state friction angle φ_c .

The dependent quantity was the support force F or the pressure on the piston, $p = F/A_{\text{piston}}$. Time was not considered because the behaviour of dry sand was assumed rate-independent. Also the stiffness was not taken into account, because absolute displacements played a minor role for the investigation of the collapse load.

Those quantities were combined to the following dimensionless variables:

- normalised support pressure $\dfrac{p}{\gamma_d\, D} = \dfrac{F}{(\pi\, D^2/4)(\gamma_d\, D)}$,
- normalised piston advance s/D ,
- cover-to-diameter ratio C/D ,

- density index I_d ,
- critical state friction angle φ_c .

In analogy to the bearing capacity formula, the normalised support pressure *at failure* for cohesionless soil can be expressed as

$$N_D = \frac{p_f}{\gamma_d \, D} \quad , \tag{3.24}$$

with the support pressure at failure p_f.

Necessary support pressures can be transferred from model to prototype scale, i.e. $(N_D)_{\text{model}} = (N_D)_{\text{prototype}}$, if the following relations are fulfilled:

$$\left(\frac{C}{D}\right)_{\text{model}} = \left(\frac{C}{D}\right)_{\text{prototype}} \quad , \tag{3.25}$$

$$(I_d)_{\text{model}} = (I_d)_{\text{prototype}} \quad , \tag{3.26}$$

$$(\varphi_c)_{\text{model}} = (\varphi_c)_{\text{prototype}} \quad . \tag{3.27}$$

Without exception, all (theoretical and numerical) models need strength parameters of the soil for the prediction of N_D. In nearly all cases a friction angle φ is required. However, the determination of strength parameters on low stress levels is difficult. As will be outlined in Sec. 3.4, φ_c seems to be suited much better for the description of the problem than φ_p: φ_c can be considered independent of stress level and initial density; moreover, the load-displacement curves reach a sort of *critical state*.

By means of the presented dimensionless quantities, results from the author's experiments can be compared with theoretical predictions and to similar tests that were performed in the centrifuge. Also, a parametric study on the model scale can reveal which variables do influence the resulting N_D and which do not.

Evaluation method

In addition to uncertainties from the force correction ΔF, the force readings showed a reasonable amount of scatter. To quantify the necessary support pressure N_D, all load-displacement curves were evaluated as shown in Fig. 3.43: an interval was defined manually, in which the force measurements reached a residual value. This interval was considered as range for necessary support pressure N_D. The mentioned uncertainties only allow to quantify N_D as a range for each test.

In the following, the *mean values* for a correction with $\Delta F = (\Delta F_1 + \Delta F_2)/2$ serve to illustrate the general soil behaviour. For the sake of clarity the bounds $F1$ and $F2$ are not plotted.

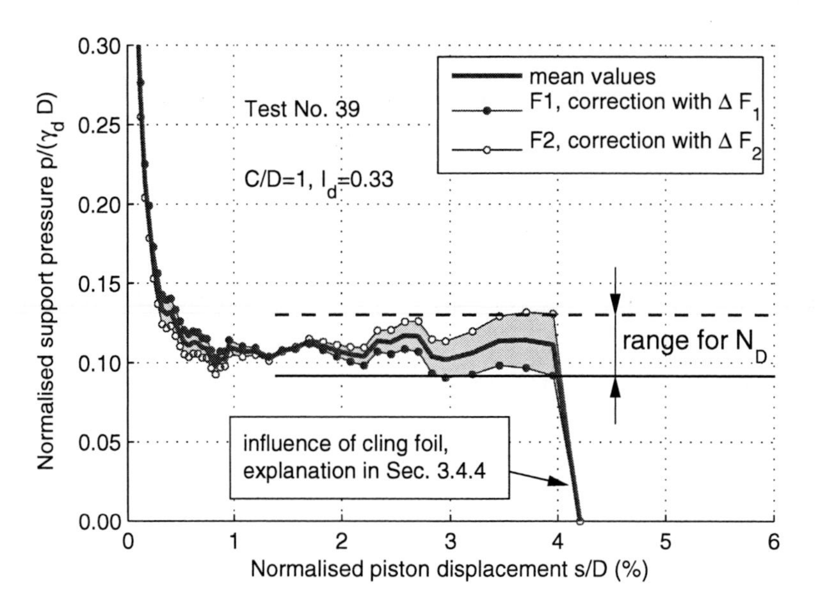

Figure 3.43: Load displacement curve for $C/D = 1.0$ (test No. 39)

Load-displacement curves

The obtained load-displacement curves of loose and dense samples for a cover-to-diameter ratio $C/D = 1.0$ are shown in Fig. 3.44. The force readings for very small displacement show a little scatter, because they were certainly influenced by the contact preparation with the load cell. Fig. 3.45 illustrates the difference between the force-displacement behaviour of loose and dense samples. The "dense" curves dropped steeply to a relatively low value as compared with the "loose" curves. But with continuing displacements, the resultant force on the piston in the dense samples increased again, reaching the same residual value for both curves after relative displacements of 2 to 3 %.

The force readings dropped to zero after different advance steps for each test.[13]

[13] Otherwise, the tests were stopped at relative displacements of approx. 6 ... 7%.

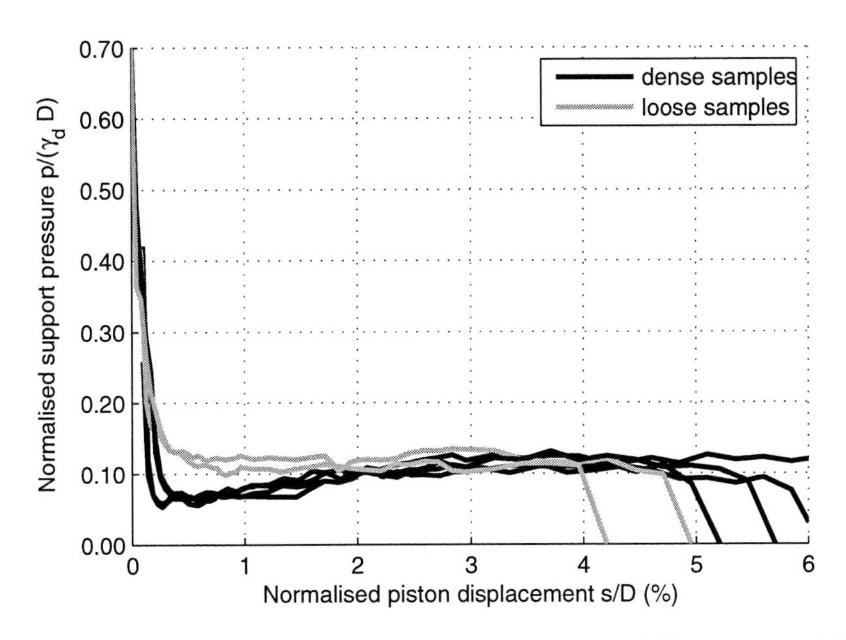

Figure 3.44: Development of normalised support pressure for $C/D = 1.0$, material $S1$

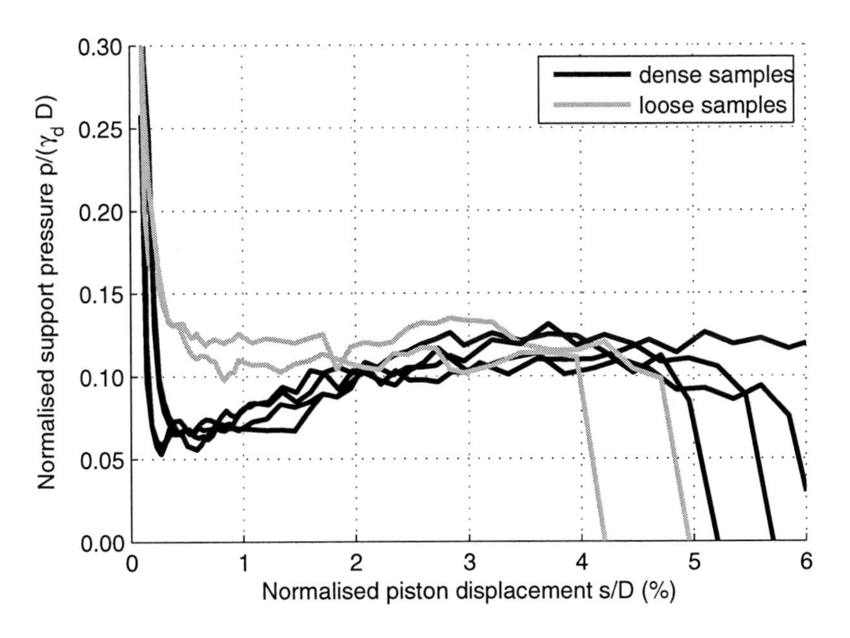

Figure 3.45: Detail of Fig. 3.44

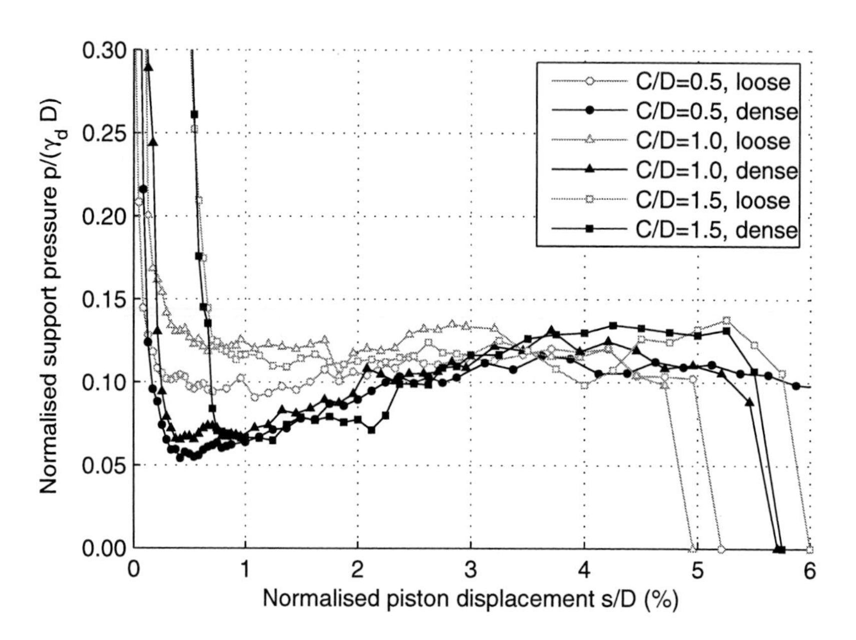

Figure 3.46: Development of necessary support pressure for different C/D ratios [14]

As illustrated in Fig. 3.46 for tests with material $S1$, the normalised support pressures for different overburdens tended to the same residual value. In other words, the influence of C/D on the residual support pressure fell within the accuracy of the evaluation method.

Necessary support pressure

Tab. 3.5 shows all results for the second series of experiments, indicating the obtained range for N_D. All load-displacement plots for the 52 tests are compiled in App. B.

No.	C/D	mat.	e	I_d	range for N_D	
01	0.25	S1	0.49	0.79	0.09	0.15
02	0.25	S1	0.48	0.81	0.10	0.13
03	0.25	S1	0.68	0.22	0.11	0.15
04	0.25	S1	0.64	0.35	0.10	0.13
05	0.25	S2	0.64	0.73	0.10	0.12
06	0.25	S2	0.64	0.76	0.09	0.12

[14]As protection measure for the load cell, the experiments for $C/D > 1.0$ started with an initial displacement of 0.5 mm. Therefore the curves seem shifted to the right.

07	0.25	S2	0.85	0.07	0.11	0.15
08	0.25	S2	0.84	0.09	0.09	0.12
09	0.50	S1	0.52	0.70	0.09	0.13
10	0.50	S1	0.51	0.74	0.08	0.11
11	0.50	S1	0.51	0.74	0.09	0.13
12	0.50	S1	0.47	0.84	0.09	0.13
13	0.50	S1	0.67	0.25	0.09	0.14
14	0.50	S1	0.64	0.34	0.10	0.13
15	0.50	S1	0.64	0.35	0.09	0.12
16	0.50	S1	0.64	0.35	0.10	0.13
17	0.50	S1	0.64	0.35	0.09	0.12
18	0.50	S2	0.66	0.69	0.10	0.14
19	0.50	S2	0.66	0.69	0.11	0.14
20	0.50	S2	0.65	0.71	0.07	0.10
21	0.50	S2	0.64	0.74	0.07	0.12
22	0.50	S2	0.88	-0.01	0.13	0.17
23	0.50	S2	0.87	0.01	0.13	0.17
24	0.50	S2	0.87	0.01	0.13	0.17
25	0.50	S2	0.87	0.01	0.14	0.18
26	0.50	S2	0.84	0.10	0.11	0.15
27	0.75	S1	0.49	0.79	0.08	0.12
28	0.75	S1	0.48	0.83	0.11	0.14
29	0.75	S1	0.66	0.29	0.10	0.14
30	0.75	S1	0.66	0.29	0.10	0.14
31	0.75	S2	0.66	0.69	0.11	0.15
32	0.75	S2	0.62	0.82	0.09	0.14
33	0.75	S2	0.88	-0.03	0.10	0.14
34	0.75	S2	0.84	0.11	0.11	0.15
35	1.00	S1	0.52	0.70	0.10	0.14
36	1.00	S1	0.51	0.73	0.08	0.12
37	1.00	S1	0.51	0.73	0.09	0.14
38	1.00	S1	0.48	0.82	0.10	0.14
39	1.00	S1	0.64	0.33	0.09	0.14
40	1.00	S1	0.64	0.33	0.10	0.14
41	1.00	S2	0.64	0.73	0.08	0.13
42	1.00	S2	0.64	0.73	0.09	0.13
43	1.00	S2	0.85	0.08	0.11	0.15
44	1.00	S2	0.84	0.09	0.09	0.14
45	1.00	S1	0.50	0.77	0.11	0.16

46	1.00	S1	0.49	0.80	0.10	0.15
47	1.00	S1	0.67	0.25	0.09	0.13
48	1.00	S1	0.65	0.29	0.10	0.14
49	2.00	S1	0.50	0.76	0.11	0.17
50	2.00	S1	0.50	0.76	0.11	0.17
51	2.00	S1	0.68	0.22	0.10	0.16
52	2.00	S1	0.65	0.29	0.11	0.16

Table 3.5: Results for the necessary dimensionless support pressure

3.3.4 Interpretation of results

The residual level of the normalised support pressure is interpreted as (dimensionless) necessary support pressure N_D: a lower pressure (in a pressure-controlled test or in the pressure chamber of a shield machine) would lead to infinite displacements, i.e. the collapse of the tunnel.

Influence of cling foil sealing

The influence of the cling foil becomes obvious as sharp pressure drop at the end of each test. The drop did not always occur at the same advance step, which leads to the conclusion that the drop was *not* due to friction in the apparatus, but depended on the cling foil configuration during preparation of the test. Moreover, the rising branch of force readings for nearly all dense samples suggests that the cling foil did not constrict the load transfer from soil to piston.

Influence of (relative) density

The generic shapes of the load-displacement curves resemble the behaviour of sand in a shear test (cf. Fig. 3.53). After the peak in the stress-strain curve, the dense sand shows softening and loses strength. On the contrary, loose samples generally do not experience a peak, but reach their strength monotonically.

There are two explanations for the "dense" load-displacement curves in the performed experiments:

- In the PIV analyses a failure mechanism was detected that consists of a sliding wedge and an overlying chimney. For an advance step of $s/D = 2.0$ % an arch can be observed. This arch redirects the weight above the sliding wedge onto the tunnel lining and the surrounding ground in such a way that the soil in front of the piston face builds up less pressure on the face.

- When the sliding wedge has formed, the shearing resistance of the sand influences the pressure on the tunnel face. As the shearing resistance of a dense sand decreases after the peak (*strain softening*), the force on the piston increases again.

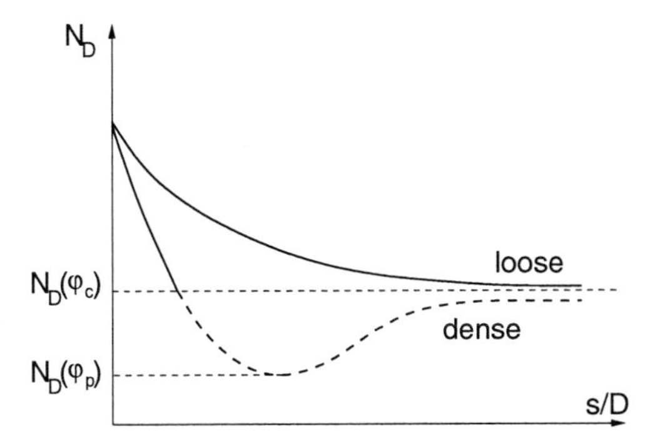

Figure 3.47: Interpretation of "dense" and "loose" curves

The fact that "dense" and "loose" curves reach the same residual value (Fig. 3.47) suggests that the failure process is eventually governed by φ_c. A necessary support pressure $N_D(\varphi_c)$ which is determined using φ_c (rather than φ_p) is always a safe estimate for a dense soil, also when large strains are reached. This interpretation overcomes some of the difficulties related to the transfer of results from model to prototype scale: conclusions from the model tests for *loose* sand (and *dense* sand at large strains) equally hold on the prototype scale, because φ_c can be considered independent of stress level (cf. Sec. 3.4).

Consideration of φ_p is only meaningful up to the peak strength; if larger strains cannot be ruled out, $N_D(\varphi_p)$ is unsafe.

Influence of overburden

Fig. 3.48 summarises the results graphically. It shows the ranges for N_D for all 52 tests as bars. The bottom shows two plots of N_D vs. C/D, one for each material. The evaluation for material $S2$ shows some scatter. Anyhow, there seems to be no influence of overburden on N_D, neither for $S1$ nor $S2$. This is supported by experimental evidence by *Léca* and *Dormieux* [77] and numerical results by *Ruse* [107].

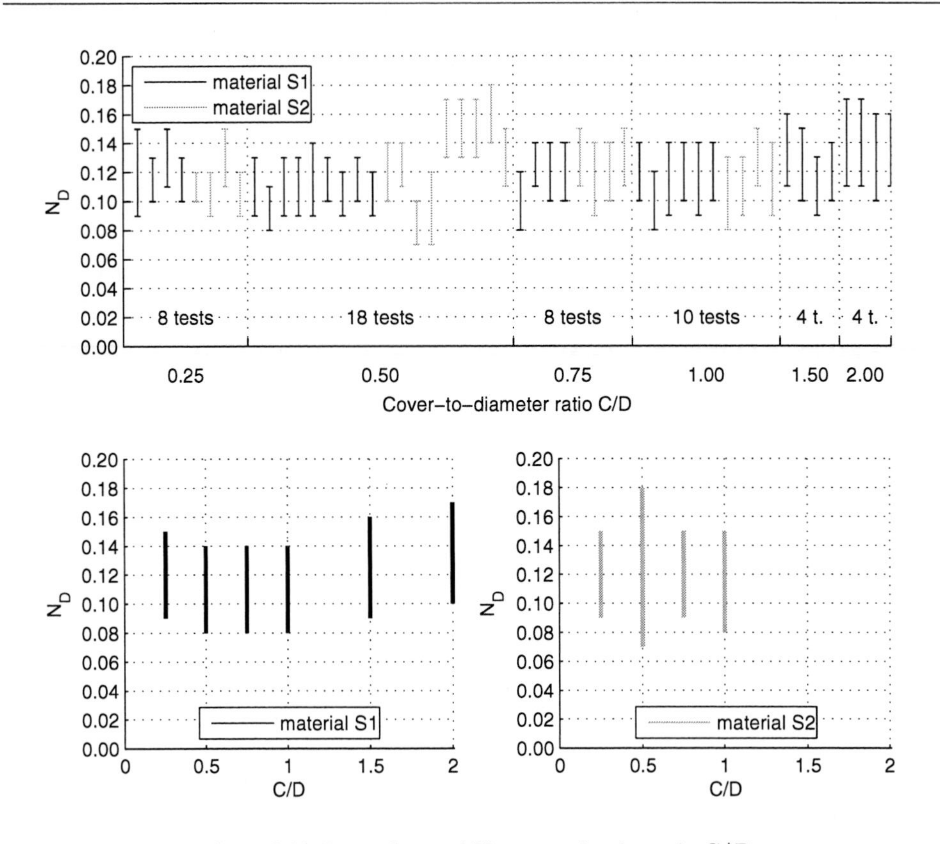

Figure 3.48: Dependence of N_D on overburden ratio C/D

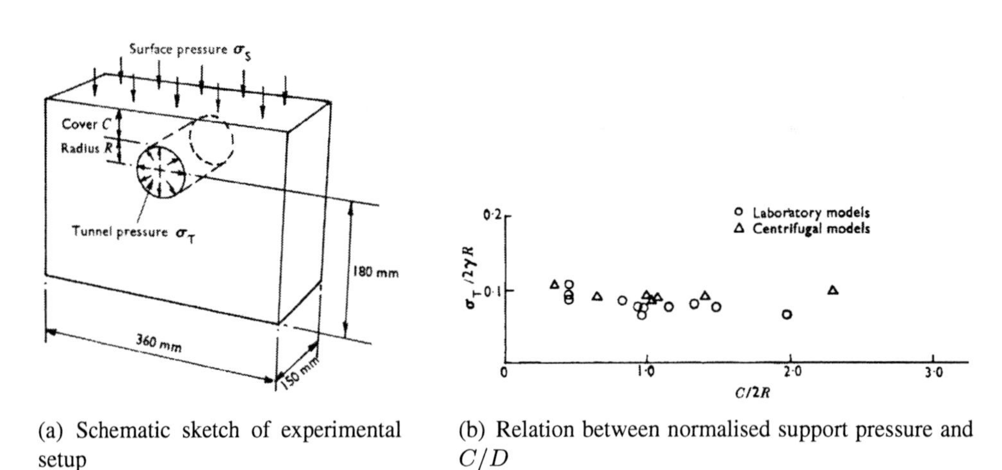

(a) Schematic sketch of experimental setup

(b) Relation between normalised support pressure and C/D

Figure 3.49: Experimental setup and results by *Atkinson* and *Potts* for the plane strain problem, [7]

Also, for the plane strain problem (Fig. 3.49 a), *Atkinson* and *Potts* [7] found no influence of C/D on the normalised support pressure, neither for ng- nor 1g-conditions (Fig. 3.49 b).

Influence of stress level

The author's results were put into context with published stress-strain curves from *Chambon* and *Corté* [24] and *Plekkenpol* et al. [101], who conducted centrifuge tests of the same problem with comparable geometry and soil properties (cf. Chap. 2).

Chambon and *Corté* performed tests at an acceleration of 50g. They used fine Fontainebleau sand with d_{50} = 0.17 mm at relative densities between 0.65 and 0.92. The mentioned friction angle in the range between 38° and 42° seems to be a peak friction angle.

Plekkenpol et al. conducted their centrifuge tests at an acceleration of approximately 75g. They used fine sand (d_{50} = 0.16 mm) with relative densities between 0.47 and 0.99, quoting (peak) friction angles between 34.5° and 44°.

To compare the shape of the published curve with the author's observations, the test results are plotted as normalised support pressure $p/(\gamma_d\, D)$ vs. normalised displacement s/D in Fig. 3.50.

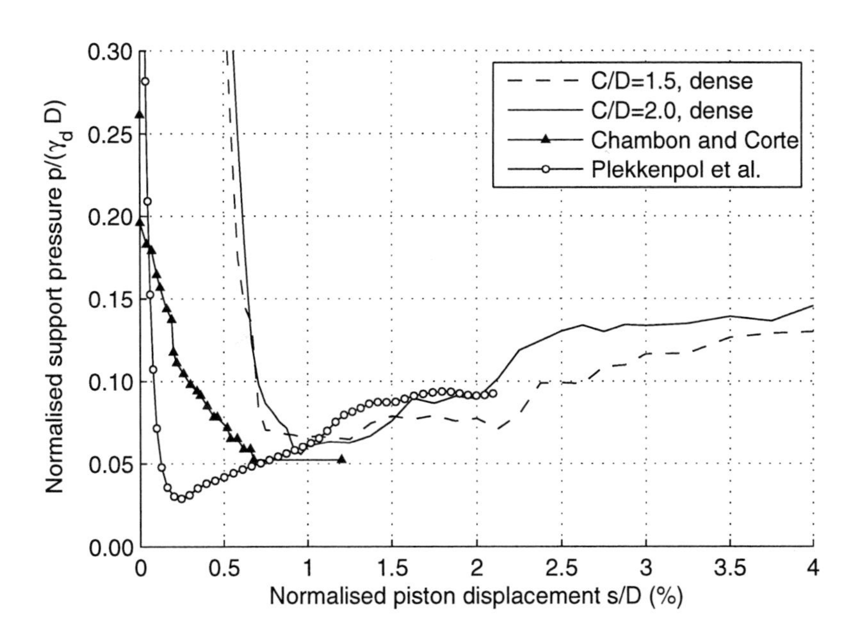

Figure 3.50: Comparison of 1g sandbox and centrifuge test results

The shape of the curves obtained in the 1g tests is in good agreement with results of the centrifuge experiments. There is a remarkable coincidence in terms of pressure level *and* relative displacement between the measured curves from centrifuge and 1g tests. As the tests by *Chambon* and *Corté* were pressure-controlled, they possibly missed the rising branch of the (dense) load-displacement curve.

If the necessary support pressure N_D is dependent on the critical state friction angle φ_c *alone* and φ_c can be assumed independent of stress level, then additional tests at elevated stress levels are not necessary for the author's line of argumentation.

Influence of sand type

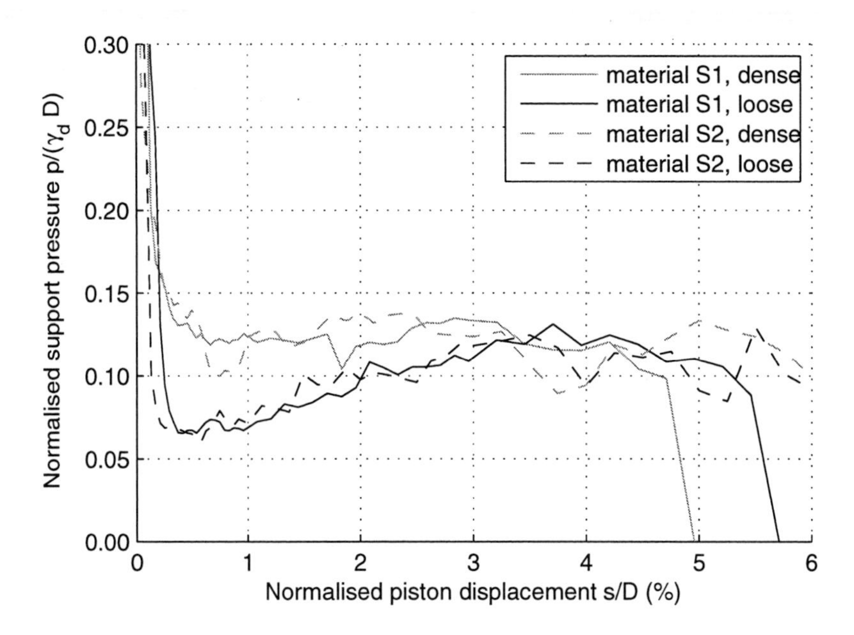

Figure 3.51: Development of necessary support pressure for different materials for $C/D = 1.0$

Fig. 3.51 shows examples for the behaviour of materials $S1$ and $S2$ with comparable density in the sandbox tests. It can be concluded that there is no significant influence of grain size distribution on the residual earth pressure on the piston. This observation is supported by an overview of all test results (Fig. 3.48), which indicates that the difference in N_D for both materials were marginal. In terms of necessary support pressure, the scaling effect is, therefore, smaller than the uncertainty of the evaluation method.

Trapdoor analogy

Again, there is a strong connection of the load-displacement measurements to the trapdoor problem. *Terzaghi* recorded the support force underneath the trapdoor for

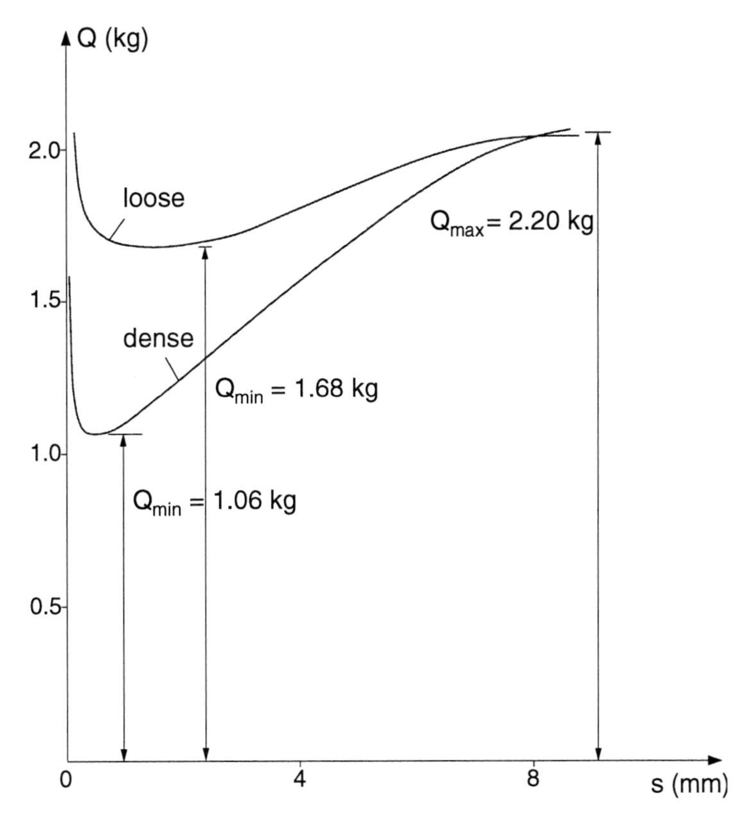

Figure 3.52: Relation between the vertical displacement s of trapdoor and support force Q, [73] (measurements reported by *Terzaghi*)

the active case. His results are reproduced in Fig. 3.52[15] and reveal the same behaviour as the author's tests. The curves have been reproduced by others, cf. e.g. *Papamichos* et al. [100] or *Dewoolkar* et al. [32]. This supports the idea that not only the kinematics of the problem can be connected to the trapdoor problem, but also the support forces.

3.4 Strength parameters for sand at low stress levels

To compare theoretical and numerical results for the necessary support pressure with the author's measurements, the strength parameters of the used material must be known. These are also required as input parameters for the numerical model (Chap. 4). Therefore, this section outlines how the author quantified shear strength parameters for the applied materials at low stress levels in the order of a few kilopascals.

[15] *Széchy* [120] attributes the same measurements to *Künzl*.

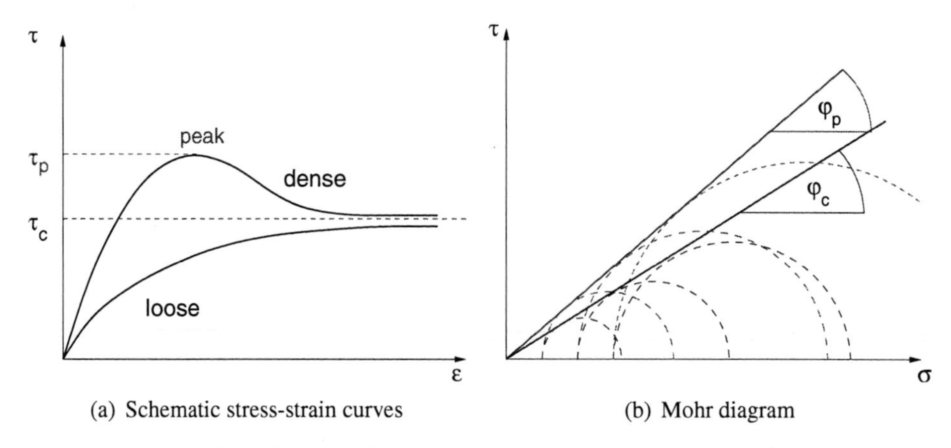

(a) Schematic stress-strain curves (b) Mohr diagram

Figure 3.53: Idealised behaviour of dry sand in a shear test

Idealised stress-strain curves for sand (Fig. 3.53) indicate that the strength of dense sand can be expressed in terms of the peak friction angle φ_p. The critical state friction angle φ_c is indicative for the strength of sand at large shear strains, no matter if the sand was initially dense or loose. It is assumed that dry sand does not have any shear strength at zero stress. Therefore, cohesion c is not considered in the following.

Critical state friction angle φ_c

The critical state friction angle φ_c characterises the strength of sand for large shear strains. It can be determined experimentally in undrained triaxial tests, preferably with loose samples. Drained triaxial test are also possible. These require much larger shear strains, though (*Herle* [51]). Large shear displacements are also achieved in ring shear tests.

In contrast to φ_p, the critical state friction angle φ_c is considered independent of initial density and stress level. *Been* et al. [14] found a small decrease of φ_c for very low stress levels. They pointed out, though, that these observations do not lead to a general rejection of the above assumption. For the applied sand $S1$, *Laudahn* [76] performed a series of triaxial compression tests at lateral stress levels between 100 and 2000 kPa with varying initial densities. As expected, he did not find a pronounced influence of stress level and density on φ_c. Other experimental studies with sand support these statements (e.g. *Verdugo* and *Ishihara* [129] or *Ishihara* [56]).

Rainer and *Fellin* [104] investigated φ_c of a fine grained sand in a ring shear apparatus. They, also, did not find an influence of stress level on the critical state friction angle.

Miura et al. [89] developed the following test procedure to quantify φ_c: a pile of sand is poured onto a circular base consisting of an inner core and an outer ring.

(a) Preparation of sand pile

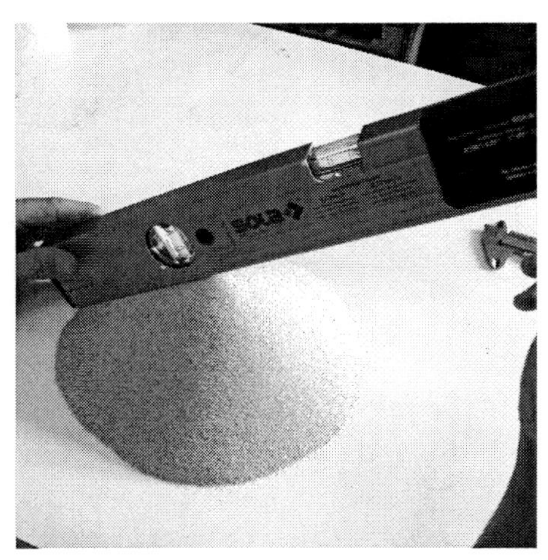

(b) Measurement of slope angle

Figure 3.54: Illustration of sand pile experiments, courtesy of *D. Renk*

Then, after slowly lowering the outer ring, a cone of sand remains on the inner core. The slope of the remaining cone of sand can be interpreted as critical state friction angle φ_c.

The angle of repose of a loose tip of dry soil subjected to toe excavation also serves as approximation to the critical state friction angle φ_c (*Bolton* [17], *Cornforth* [28]). *Herle* [51] suggested to slowly pour a pile of sand from a funnel and measure the slope of the pile (Fig. 3.54). This method was used to investigate materials $S1$ and $S2$.

An average angle of repose of $\varphi_{c,S1}$ = 32.5° (from n = 44 measurements, standard deviation $s_{\varphi_{c,S1}}$ = 1.1°) was obtained for material $S1$. For material $S2$ a value $\varphi_{c,S2}$ = 31.3° was obtained (n = 20, $s_{\varphi_{c,S2}}$ = 1.4°).

In addition, the author performed sand chute experiments to quantify φ (Fig. 3.55 a). One side of the chute's bottom was pin jointed to the base; via a string on the opposite side the chute could be tilted. The chute was filled with sand and then compacted by hand tamping with different effort to test the material at different initial densities. The height of the initially horizontal sand layer was approximately 2.3 cm.

Then the chute was slowly tilted until the material slowly slid down.[16] The inclination of the chute at this stage, α_f was recorded. The inclination of the remaining heap of sand inside the chute was interpreted as critical state friction angle φ_c of the material (Fig. 3.55 b). This procedure had the advantage that very large shear strains were achieved, which is crucial for the determination of φ_c, especially at low relative densities (*De Beer* [30]).

[16]This process can be considered a slow granular flow.

Test No.	measured angle of repose (in °) at point			
	90°	**180°**	**270°**	**360°**
1	31.9	34.5	32.4	32.6
2	33.4	33.6	32.6	32.0
3	31.3	33.3	32.3	34.2
4	32.1	33.2	34.2	31.7
5	31.7	33.7	30.1	33.0
6	31.3	33.4	33.1	31.7
7	33.5	32.2	32.0	32.8
8	31.7	31.0	33.0	32.4
9	33.6	34.1	32.2	33.6
10	30.4	34.3	31.1	32.2
11	31.2	33.2	30.5	33.2
	mean value $\varphi_{c,S1} = 32.5°$			

Table 3.6: Angle of repose for material $S1$, [76] (tests 1–6) and courtesy of *D. Renk* (tests 7–11)

Test No.	measured angle of repose (in °) at point			
	90°	**180°**	**270°**	**360°**
1	31.0	32.7	30.9	30.5
2	32.2	30.7	32.2	30.2
3	30.5	30.0	30.5	30.1
4	32.3	34.1	30.1	33.8
5	29.6	29.6	33.9	31.0
	mean value $\varphi_{c,S2} = 31.3°$			

Table 3.7: Angle of repose for material $S2$, courtesy of *D. Renk*

The results of n = 20 measurements revealed $\varphi_{c,S1} \approx 31.6$ ° (standard deviation $s_{\varphi_{c,S1}}$ = 1.0°) for material $S1$, and $\varphi_{c,S2} \approx 30.8$ ° (standard deviation $s_{\varphi_{c,S2}}$ = 0.9°, n = 23) for material $S2$.

As result of above sand pile investigations and sand chute experiments, the critical state friction angle for material $S1$ was quantified to $\varphi_{c,S1} = 32°$ and for material $S2$ to $\varphi_{c,S2} = 31°$.

Peak friction angle φ_p

Bolton [17] compiled (triaxial and plane strain) test results for sands at different confining pressures p and relative densities I_d. He pointed out the importance of taking stress level and soil density into account when determining peak strength parameters. Other influence factors that will not be taken into account in this study are grain shape [30], grain mineralogy [17] and loading history [122].

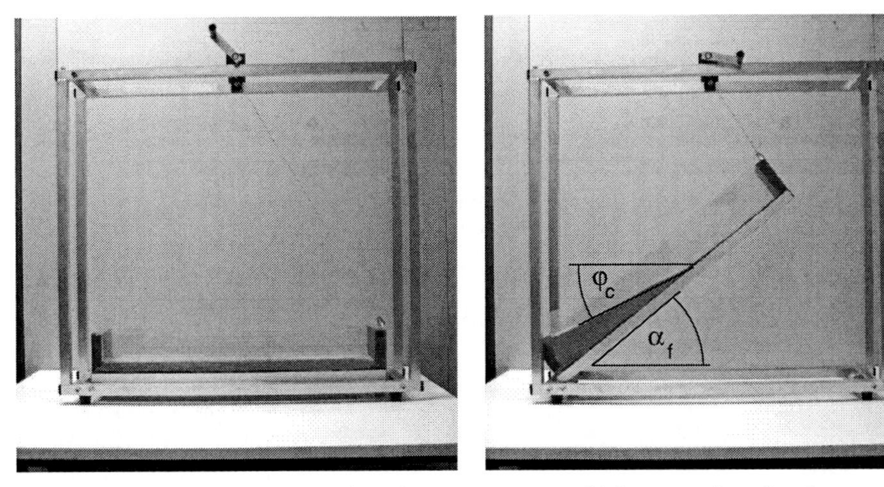

(a) Sand chute for the determination of φ_c (b) Interpretation of angles

Figure 3.55: Setup of sand chute experiments

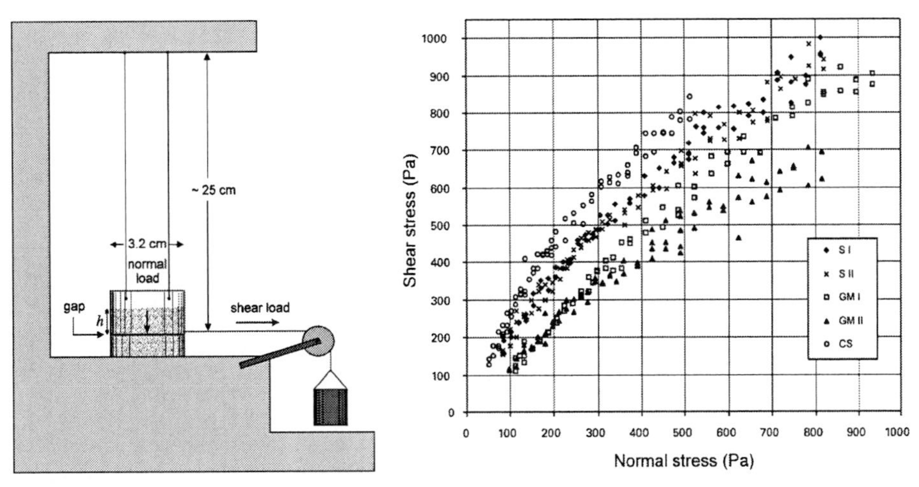

(a) Sketch of shear test apparatus (b) Shear test results for dry sand (S), glass microspheres (GM) and caster sugar (CS)

Figure 3.56: Experimental setup and results by *Schellart*, [108]

The stress levels in the 1g model tests were very low ($\approx$ 2 kPa); unfortunately, there are no appropriate standard soil mechanics laboratory tests to determine the strength parameters of the sand at such low stress level. *Schellart* [108] investigated the shear strength of fine sand (label *S* in Fig. 3.56 b) in a stress range between 0.05 and 0.9 kPa (!), making use of a *Hubbert-type* shear apparatus illustrated in Fig. 3.56 a. It consisted of an upper ring, suspended above a fixed lower ring with a gap of 0.3 mm. The rings were filled with material up to a height h above the gap; only the self-weight of the material served as normal pressure in the shear plane. Then an increasing shear load was applied via a pulley and an empty bucket, which was subsequently filled with sand. Failure was defined as a sudden horizontal displacement of the upper ring. The self-weight of the soil in the bucket at this stage was used to calculate the shear stress at failure.

Results of *Schellart*'s tests are shown in Fig. 3.56 b. The data support the hypothesis of a curved failure envelope for normal stresses below, say, 0.4 kPa. The derived friction angles for a Mohr-Coulomb failure criterion range from $\varphi \approx 65°$ for stresses below 0.1 kPa to $\varphi \approx 40°$ for stresses above 0.6 kPa.

Unfortunately, *Schellart* did not provide information on the samples' densities. As the material was poured from a height of 10 cm, the author assumes that they were rather (medium) dense than loose.

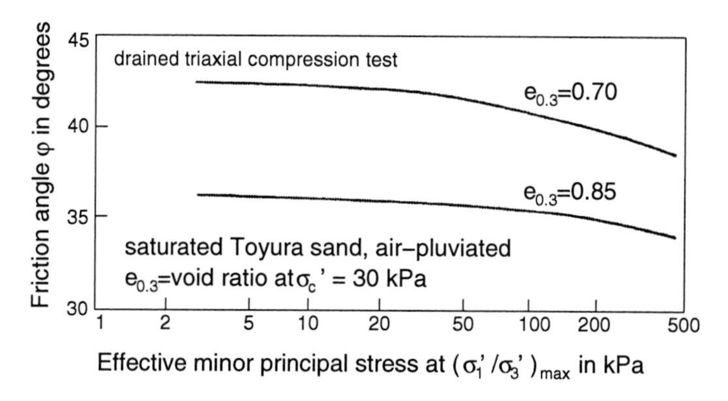

Figure 3.57: Influence of stress level and void ratio on peak friction angle φ_p for Toyoura sand, [41]

Triaxial tests at stress levels as low as 5 kPa have been performed by *Fukushima* and *Tatsuoka* [41] on Toyoura sand. These tests required a highly sophisticated test procedure and various stress corrections, such as membrane corrections, to obtain reliable results. The experimental results (Fig. 3.57) indicate that φ_p did *not* depend on confining pressure for low stress regimes ($\sigma_3 \leq$20 kPa).[17] But density I_d had a pronounced influence on φ_p. *De Beer* [30] quoted test results for different sands

[17] There is some debate on this topic, see e.g. *Cabarkapa* [22].

which reveal a difference in friction angle $\Delta\varphi_p$ of up to 15° between $I_d = 0$ and $I_d = 1$.

The author performed above mentioned sand chute experiments to investigate the influence of density on the inclination of the chute at failure, α_f, at very low stress levels. The observed inclination is *not* indicative for the *peak* friction angle, φ_p, because the dense samples failed progressively. In some cases even a single sand grain, tumbling over the surface, triggered the failure of the slope inside the chute. In the author's opinion, the inclination α_f at this point is, therefore, only representative for some intermediate friction angle between φ_c and φ_p. Nevertheless, the influence of I_d on α_f could be studied.

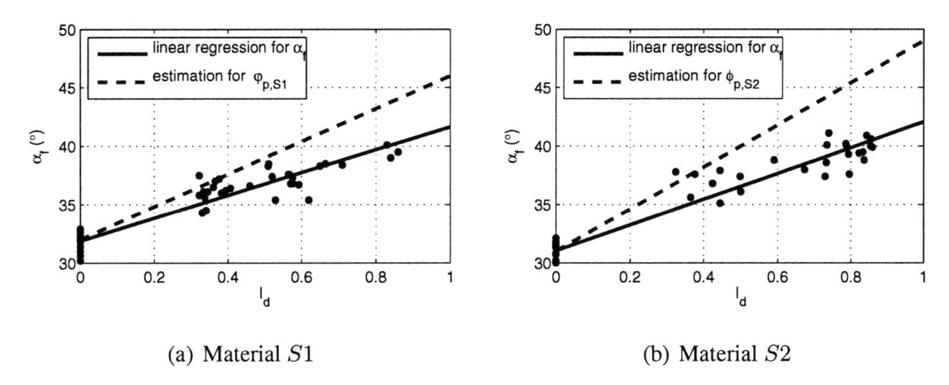

(a) Material $S1$ (b) Material $S2$

Figure 3.58: Sand chute results with linear regression and estimations for the relation between φ_p and I_d

The results of the experiments are displayed in Fig. 3.58 in a plot of inclination α_f vs. density index I_d. It was assumed that the (critical) void ratio e_c can be approximated by the maximum void ratio $e_{\max}$ (cf. *Herle* [51]). Therefore, the φ_c values are plotted vs. $I_d = 0$ in Fig. 3.58.

It becomes obvious that α_f increases with growing I_d. Both materials, $S1$ and $S2$, revealed values for α_f of up to 40° for a dense packing, thus indicating that the peak friction angles are even higher. The author guessed a relation between φ_p and I_d as a reasonable upper bound to all measurements with $I_d > 0$. In accordance with results presented by *Bolton* [17], *Fukushima* and *Tatsuoka* [41], *De Beer* [30] and *Wu* [139] a linear relationship between φ_p and I_d was adopted. For the used materials, the following relations were assumed:

$$\varphi_{p,S1} = 32^\circ + 14^\circ\, I_d \tag{3.28}$$

$$= \varphi_{c,S1} + 14^\circ\, I_d \quad , \tag{3.29}$$

and

$$\varphi_{p,S2} = 31^\circ + 18^\circ\, I_d \qquad (3.30)$$
$$= \varphi_{c,S2} + 18^\circ\, I_d \quad . \qquad (3.31)$$

The difference $\Delta\varphi_p$ between I_d = 0 and I_d = 1 is in accordance with results by *De Beer* [30]. The increase of φ_p with I_d is supported by a lot of experimental evidence at higher confining pressures (cf. *Bolton* [17]) and was termed *pyknotropy* by *Kolymbas* [74].

Dilation angle ψ

There is a lot of experimental evidence, e.g. from (drained) biaxial, triaxial or direct shear tests, that soil samples change their volume when sheared. An increase in volume is generally termed *dilation*, a decrease *contraction*.

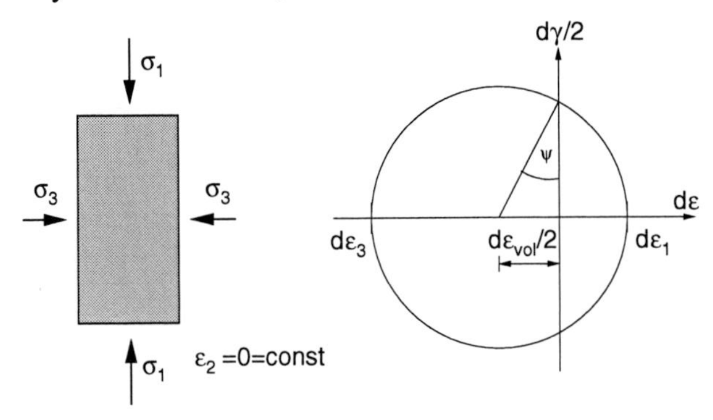

Figure 3.59: Idealised plane stain compression test with Mohr's circle of strain increments (compression positive)

In the context of plane strain compression tests (Fig. 3.59) the dilation angle ψ has been introduced as ratio of incremental volumetric to incremental shear strain

$$\sin\psi = -\frac{d\varepsilon_{\text{vol}}}{d\gamma_{13}} \quad , \qquad (3.32)$$

which can also be expressed in terms of principal strain increments as

$$\sin\psi = -\frac{d\varepsilon_1 + d\varepsilon_3}{d\varepsilon_1 - d\varepsilon_3} \quad . \qquad (3.33)$$

The peak *strength* of a sand in any drained shear test is usually associated with the maximum rate of dilation $b_p := (-d\varepsilon_{\text{vol}}/d\varepsilon_1)_p$. The dilation angle at peak strength is defined as

$$\sin\psi_p = -\frac{(d\varepsilon_1/d\varepsilon_3)_p + 1}{(d\varepsilon_1/d\varepsilon_3)_p - 1} \quad . \qquad (3.34)$$

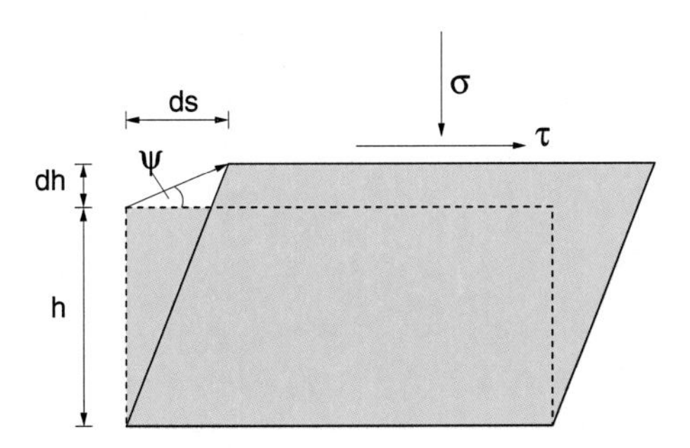

Figure 3.60: Idealised simple shear deformation, [74]

A theory by *Taylor* [123] considers (residual) friction and dilation as the two components which contribute to the peak strength of a material. For the example of simple shear (Fig. 3.60), the incremental work $dW = \tau\ ds$ can be expressed as $dW = \sigma\ \tan\varphi_p\ ds$ at the peak. If friction and dilation are considered to be independent, dW can be formulated in terms of φ_c and ψ:

$$\begin{aligned} dW = \sigma \cdot \tan\varphi_p \cdot ds &= \sigma \cdot \tan\varphi_c \cdot ds + \sigma \cdot dh \\ &= \sigma \cdot \tan\varphi_c \cdot ds + \sigma \cdot \tan\psi_p \cdot ds \quad , \end{aligned}$$

which finally yields

$$\tan\varphi_p = \tan\varphi_c + \tan\psi_p \quad . \tag{3.35}$$

Shear box tests at low stress levels with reconstituted sand by *Fannin* et al. [38] revealed a slightly different relation,

$$\varphi_p = \varphi_c + \psi_p \quad , \tag{3.36}$$

which is supported by triaxial test results by *Fukushima* and *Tatsuoka* [41].

For very loose samples and/or high stress levels ψ_p is zero.[18] But for dense samples at lower stress levels $\psi_p \neq 0$. It has been shown that ψ_p depends on stress level and, to a greater extent, initial density of the sample [122].

Combining (3.36) with (3.29) and (3.31) yields approximate relations between dilation angle and density index for low stress levels:

$$\psi_{S1} = 14^\circ\ I_d \quad , \tag{3.37}$$

and

$$\psi_{S2} = 18^\circ\ I_d \quad . \tag{3.38}$$

[18] In other words, the sample does not exhibit any increase in volume when sheared.

3.5 Critical evaluation of qualitative and quantitative results

Comparison with analytical predictions

The results of the force measurements at low stress levels are compared with predictions of the various proposed models presented in Chap. 2. For this comparison an overburden $C/D = 1.0$ was chosen, cohesion was neglected. The range of experimentally obtained N_D values is plotted against the critical friction angle.

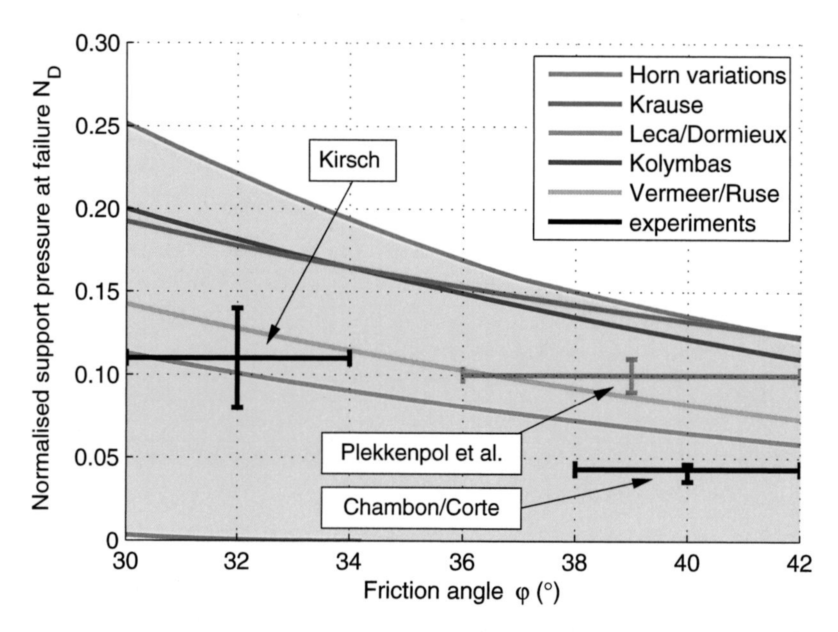

Figure 3.61: Comparison of experimental results with analytical predictions (the range of possible *Horn* predictions is shaded in red)

Fig. 3.61 shows a comparison of analytical predictions and experimental results. Predictions and measurements are of the same order of magnitude; to be more precise, the experimental results lie in the middle of the range of *Horn*-model predictions. The approaches by *Kolymbas* and *Krause* overestimate the experimental results. The upper bound solution by *Léca/Dormieux* and the empirical approach by *Vermeer/Ruse* seem to approximate the experimental observations well.

The experimental results by *Chambon* and *Corté*, which are also included in the comparison, might underpredict the residual value for N_D, most probably because the rising branch of the load-displacement curve could not be captured in their pressure-controlled tests.

Review: sensitivity analysis

In Sec. 2.5.2 the concept of sensitivity analysis was outlined as a measure for the quality of the proposed models. At this point the experimental can be combined with the statistical comparison of the models. Hence, the uncertainty of the predictions for N_D was calculated; the assumed distributions of the input parameters are indicated in Tab. 3.8.

parameter	**mean value** μ_i	**coefficient of variability** V_i	**standard deviation** σ_i	*source*
self unit weight γ	16.7 kN/m^3	4.0 %	0.67 kN/m^3	*assumption*
friction angle φ_c	32°	3.0 %	1.0°	sand chute tests
tunnel diameter D	10 cm	0.1 %	0.01 cm	*assumption*
tunnel cover C	10 cm	5.0 %	0.50 cm	*assumption*

Table 3.8: Input parameters for sensitivity analysis

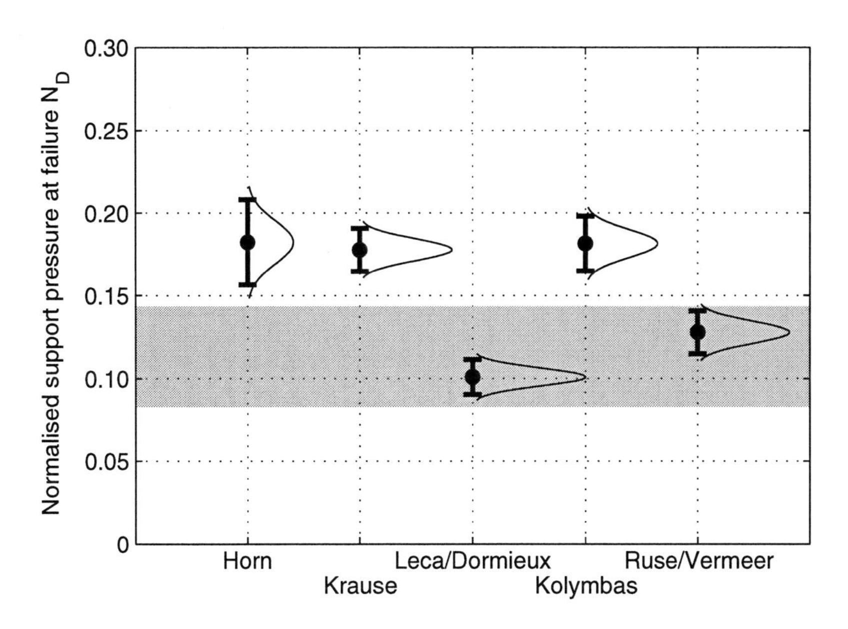

Figure 3.62: Variation of model predictions for N_D and experimental results (95% confidence interval for tests No. 35–40)

The variation of result for the chosen variation in input parameters is shown in Fig. 3.62. The bars indicate the 95% confidence intervals. The range for N_D obtained from the small-scale experiments (C/D = 1.0) is shaded in grey.

It becomes obvious that only the predictions by *Ruse/Vermeer* and *Léca/Dormieux* intersect the experimental results on a 95% confidence level. The other models in question significantly overestimate the necessary support pressure.

Relation between qualitative and quantitative results

Both test series, the PIV investigation and the force measurements, were performed under the same conditions and with the same geometric configuration. Therefore, it is possible to relate results from both test series: Fig. 3.63 shows an example for incremental displacements within the soil body at different stages throughout collapse of the face. As outlined before, the initially loose samples reveal the same displacement pattern throughout the whole test. The dense samples, in contrast, show a development of the failure zone towards the ground surface.

At a point where the "dense" load-displacement curve has its minimum, only very small soil displacements are visible. This supports the idea that the maximum strength of the soil is mobilised at this stage. After this minimum the force on the piston rises again, and the failure zone develops towards the soil surface.

The failure mechanism has not yet reached the ground surface, when the load-displacement curve reaches its residual value. According to the silo equation, the vertical stresses reach a limit value once the silo has a height of roughly D_{silo}. From this point onwards, vertical load exerted by the chimney on the sliding wedge remains constant. This explains why the measured force does not increase any further.

It should be noted that, although the plots of incremental soil displacements shown in Sec. 3.2 reveal a completely different pattern for dense and loose samples, the residual N_D is identical. This holds evenly for the the two applied sands $S1$ and $S2$.

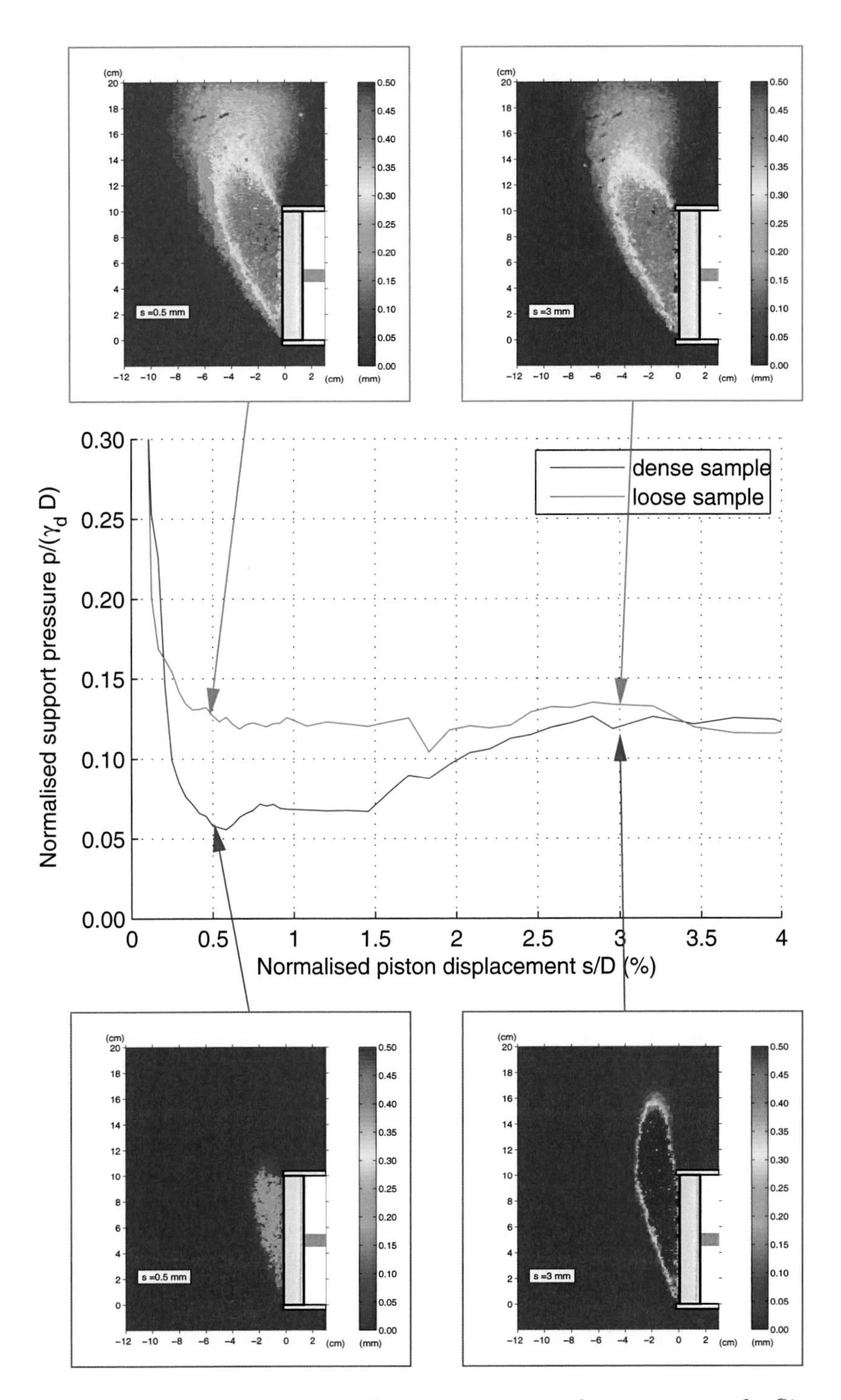

Figure 3.63: Relation between displacement pattern and support pressure for $S1$

SYNOPSIS

To assess the quality of proposed models for face stability analysis, two series of small-scale experiments at single gravity were performed. In the first series of experiments the failure mechanisms in dense and loose sand with different overburden were investigated, making use of Particle Image Velocimetry. The resulting support force on the tunnel face was studied in a second series of experiments. It was demonstrated that neither the density nor the overburden had an influence on the necessary support force. The observed failure mechanism was significantly different for different initial densities, though.

A comparison of the experimental results with predictions by several theoretical/numerical models has shown that the measured necessary support pressure is significantly overpredicted by most of the models. Only the models by Vermeer/Ruse and Léca/Dormieux were able to predict the necessary support pressure on a 95% confidence level.

Chapter 4

Finite element modelling for face stability analysis

The overview in Chap. 2 has shown how much predictions of the necessary support pressure of various analytical approaches differ for a simple example. The performed laboratory experiments (Chap. 3) indicated that only a few of the compared approaches can predict the measured support pressure on a 95% confidence level. The others significantly overestimated the necessary support. In addition, none of the proposed models yields information about the ground displacements.

But why is it so difficult to find an analytical solution to the face stability problem, which provides information on displacements and stability? Any theoretical (or numerical) model seeks a solution for 15 unknown quantities for any point in a three-dimensional continuum: displacement (3 unknowns), strain (6 unknowns) and stress (6 unknowns). The solution must fulfil the following requirements:

- **Balance equations for mass and momentum:** These equations can be combined to three scalar equilibrium equations (3 equations). The equilibrium equations have absolute validity.
- **Compatibility:** The compatibility conditions guarantee that the strain fields can be integrated to a continuous displacement field (6 equations).
- **Constitutive model:** The missing 6 equations are provided by the constitutive model, which links stresses and strains for a particular material. The constitutive model is always an approximation to real soil behaviour, and there is no measure to assess its quality in general. It is only possible to check a model's ability to reflect the behaviour of a given material under certain conditions.

  Some requirements for constitutive models have been collated by *Kolymbas* [73].
- **Initial and boundary conditions:** The solution to a boundary value problem needs to fulfil displacement and/or traction conditions on its boundary.

Most of the analytical approaches, such as closed form solutions, limit equilibrium or limit analysis (i.e. the bound theorems), violate some of the above mentioned requirements. As a matter of fact, only simple problems can be solved analytically, applying simple constitutive models (such as linear elasticity) or restricting the solution to two-dimensional cases.

Practically speaking, only numerical methods are able to solve complex problems, fulfilling to a satisfactory amount the posed requirements. They can cope with complex geometry (of both structure and ground), staged construction and non-linear and/or time-dependent material behaviour, etc. which is essential for the solution of the face stability problem.

Different strategies have been developed: some treat the ground as continuum (**F**inite **E**lement **M**ethod (FEM) and **F**inite **D**ifference **M**ethod (FDM)), some take the discontinuous nature of, e.g., fractured rock into account (**D**iscrete **E**lement **M**ethod (DEM)). In recent years the **D**iscrete **P**article **M**ethod (DPM), which models the soil grains as individual particles, has been increasingly applied in research.

A variety of numerical tools is available. *Gioda* and *Swoboda* [44] gave an extensive overview of the developments and applications of numerical methods to tunnelling problems, with a focus on finite elements.

The intention of the numerical study of this work was to check the suitability of two constitutive models, an elastoplastic and a hypoplastic one, for the following task: to predict the displacement pattern and the necessary support pressure for a face stability analysis. The author favoured a continuum approach with finite elements rather than a discrete modelling of the sand particles for the following reasons: finite element calculations have become a standard tool in engineering practice. Therefore, the results of the present study can yield recommendations for the application of readily available material models. Moreover, a one-to-one modelling of the sand in the sandbox experiments with discrete particles is still prohibited by computational limits.[1]

For all calculations the commercial software ABAQUS/Standard by *SIMULIA* was used. The performed sandbox experiments offered the opportunity to compare the simulation results of the two material models not only with each other, but also with the physical reality of the laboratory experiments.

This chapter starts with a brief outline of the finite element method. The applied constitutive models are explained, and their characteristics discussed on the basis of element test simulations, namely a drained triaxial and an oedometer test. Then, the procedure to find a suitable finite element model for the sandbox experiments

[1] At present discrete element models can (only) cope with a couple of hundred thousands of three-dimensional particles in reasonable computation times.

is explained in detail. Finally, numerical predictions with both material models are compared with the soil behaviour observed in the laboratory.

4.1 Brief introduction to finite element theory

There are many books, periodicals and conferences on finite element theory (details can e.g. be found in *Zienkiewicz* and *Taylor* [141] or *Potts* and *Zdravković* [102]). Here, only the main ideas of finite element theory will be outlined:

The virtual work formulation serves as starting point for the finite element method: the virtual work by internal and external forces must vanish for a body in equilibrium. The virtual work equations are derived from the equilibrium equations and the boundary conditions by introducing virtual displacements. The resulting virtual work is then integrated over the volume and surface of the body. The equations are not fulfilled in every point of the continuum, but only in the *weighted integral mean*. Therefore, this approach is called *weak formulation*.

To solve the governing equations, the problem geometry is discretised into elements with finite dimensions, called *finite elements*. These elements are defined by their shape (triangular or quadrilateral for 2D-elements; tetrahedral, triangular prism or brick for 3D-elements) and the number of *nodes*. These nodes are distinct points, usually on the boundary of the element, which serve as reference points for the approximation of the displacement field.

In geotechnical applications, usually the displacements of a continuum are chosen as primary variables. This means that stress and strain are treated as secondary quantities which can be derived from the displacement field. One major assumption in a finite element model is the *shape function*, which describes the variation of displacement within an element.

The virtual work equations can be solved for the nodal displacements, taking boundary conditions and the given constitutive model into account. From these, the displacement, strain and stress fields for the whole continuum can be calculated.

For linear problems, i.e. those with a linear material model and small displacements, nodal displacements and nodal forces are uniquely related via the global stiffness matrix K. This stiffness matrix is constant for each step[2] of a finite element calculation. Fig. 4.1 a illustrates how a given load ΔP is related to the corresponding displacement Δu in an analysis step.

[2]Finite element calculations are divided into **steps**, in which a certain load/displacement configuration is imposed onto the model. Each step comprises a time period t; for rate-independent problems, t serves as auxiliary variable for the solution algorithms.

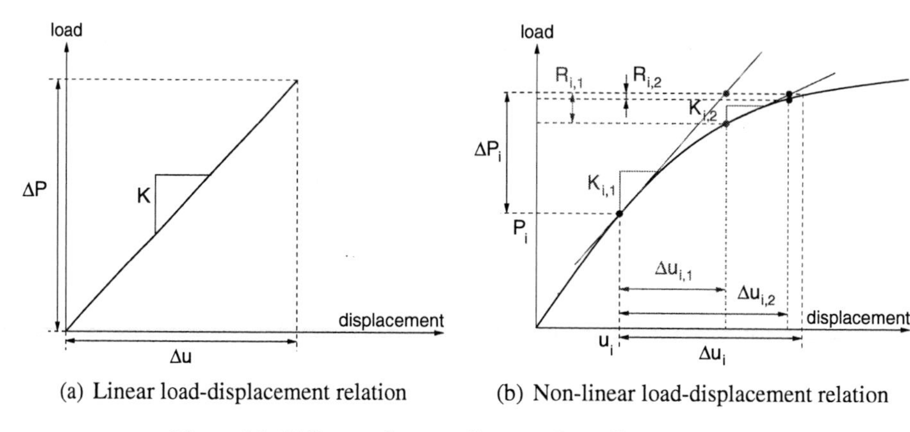

(a) Linear load-displacement relation (b) Non-linear load-displacement relation

Figure 4.1: Difference between linear and non-linear problems

For non-linear material models and/or non-linear geometry due to large displacements, the virtual work principle does not lead to a system of linear equations; in other words, the global stiffness matrix is not constant over a step. Therefore, an incremental-iterative approach is applied to solve the governing equations; i.e. the loading ΔP of a body, either by surface loads or defined displacements, must be applied in increments[3] ΔP_i (Fig. 4.1 b): the FE code guesses a corresponding displacement increment Δu_i, and the residual work R_i (i.e. the difference between virtual work of internal and external forces) is calculated. Non-linear constitutive models must be integrated numerically to calculate the work of the internal forces.

If the residual work is above a predefined tolerance, the body is not in equilibrium and iteration schemes come into action. ABAQUS uses the Newton-Raphson method, which updates the global stiffness matrix K_i in each iteration[4] of every increment. If no displacement increment Δu_i can be found, which fulfils the tolerance requirements, the governing equations did not converge. In this case, a smaller load increment ΔP_i is applied, and a second iteration cycle starts.

Some finite element programs, such as ABAQUS, allow to implement constitutive models via a *user-defined subroutine* (UMAT). For a given stress $\boldsymbol{\sigma}(t)$ and strain $\boldsymbol{\varepsilon}(t)$ at time t and given time and strain increment (Δt and $\Delta\boldsymbol{\varepsilon}$, respectively), this user-subroutine provides the stress $\boldsymbol{\sigma}(t + \Delta t)$ and the Jacobian $\partial\Delta\boldsymbol{\sigma}/\partial\Delta\boldsymbol{\varepsilon}$ at time $t + \Delta t$.[5]

[3]Each step can be subdivided into **increments**, which is necessary for non-linear problems. Every increment has a duration Δt with $\sum \Delta t = t$. The maximum number of increments per step can usually be defined by the user.

[4]In non-linear analyses, a certain number of **iterations** has to be performed until the residual work is below the given tolerance. ABAQUS allows to define the maximum number of iterations per increment.

[5]Some iteration schemes need the Jacobian to find global equilibrium.

If the iteration schemes still diverge after a number of iteration cycles, the desired loading of the body cannot be equilibrated. In other words, the body fails to carry the applied loads. Due to the small load increments and the large number of iteration cycles, the computation effort increases significantly close to limit states.

4.2 Description and calibration of the applied material models

Two constitutive models were compared by means of finite element simulations of the sandbox experiments. The author chose an elastoplastic model with Mohr-Coulomb failure criterion and a hypoplastic model. The Mohr-Coulomb model is a standard constitutive model and frequently applied for geotechnical purposes. Its performance in the given boundary value problem might, therefore, be interesting for users in engineering practice.

The more advanced hypoplastic model was developed to describe the non-linear stress-strain relations for sand. The model takes the influence of density and stress level on soil behaviour into account. These abilities seem crucial for the simulation of the laboratory experiments. Versions of hypoplasticity have been successfully implemented as user-defined models in ABAQUS, PLAXIS and TOCHNOG.

4.2.1 Mohr-Coulomb model

The Mohr-Coulomb model is a simple elastoplastic model. The difference between elastic and plastic behaviour can be illustrated in terms of idealised stress-strain curves (Fig. 4.2). While the, linear (a) or non-linear (b), elastic curves follow the same path for loading and unloading, the plastic behaviour (c) is characterised by irreversible strains.

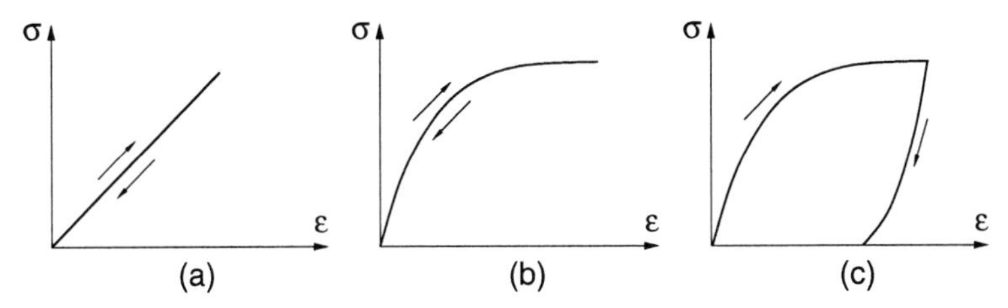

Figure 4.2: Linear elastic behaviour (a), non-linear elastic behaviour (b), plastic behaviour (c), [73]

Elastoplastic models vary in their description of the elastic and the plastic material behaviour (Fig. 4.3). Considering the one-dimensional case, let f denote the yield stress at which the behaviour switches from elastic to plastic (*yield function*): for

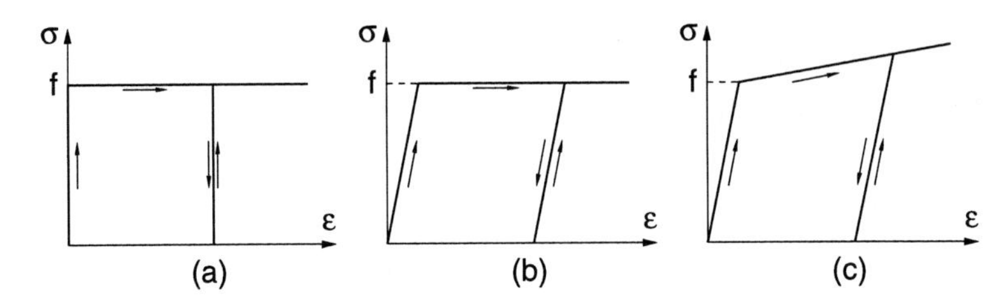

Figure 4.3: Various concepts of plastic behaviour: (a) rigid, perfectly plastic, (b) elastic, perfectly plastic, (c) elastic, plastic hardening, [73]

$\sigma < f$ the behaviour is elastic; when the stress becomes equal to f and the material is loaded further, plastic strains occur.

In three-dimensional stress and strain space the yield function cannot be formulated in terms of a simple one-dimensional inequality, because tensors are involved. In this case, the failure criterion can be formulated in terms of stress and strain invariants (or combinations of those).

An elastoplastic model requires some basic "ingredients" (cf. *Kolymbas* [72], *Potts* and *Zdravković* [102]), which help to understand the Mohr-Coulomb model:

Coaxiality The principal stress directions are assumed to coincide with the principal axes of *incremental* plastic strain.

Yield function The surface $f = 0$ separates elastic from elastoplastic behaviour. Usually, f is expressed in terms of stresses $\boldsymbol{\sigma}$ and state parameters $\boldsymbol{k}$.[6,7] For the applied Mohr-Coulomb model, the principal stresses σ_1 (major principal effective stress) and σ_3 (minor principal effective stress)[8] and strength parameters of the soil, here friction angle φ and cohesion c, are used:

$$f(\boldsymbol{\sigma}, \boldsymbol{k}) = (\sigma_1 - \sigma_3) - \sin\varphi(\sigma_1 + \sigma_3) - 2\,c\,\cos\varphi = 0 \quad . \tag{4.1}$$

This yield function can be visualised in a Mohr-diagram of shear stress vs. normal stress components (Fig. 4.4 a). In principal stress space, (4.1) plots as (irregular) hexagonal cone (Fig. 4.4 b). The hexagonal nature of the cone results from permutations of the principal stresses.

For $f(\boldsymbol{\sigma}, \boldsymbol{k}) < 0$ the material behaves elastically; stress states with $f(\boldsymbol{\sigma}, \boldsymbol{k}) > 0$ are impossible.

[6] In geotechnical engineering it is common practice to consider compressive stresses as positive. In some cases this can lead to confusion with the mechanical definition, i.e. compression negative.

[7] In general, f is formulated in terms of effective stresses. The dash is omitted, though, because no distinction is made between drained and undrained conditions (for dry sand).

[8] The intermediate principal effective stress σ_2 is not taken into consideration.

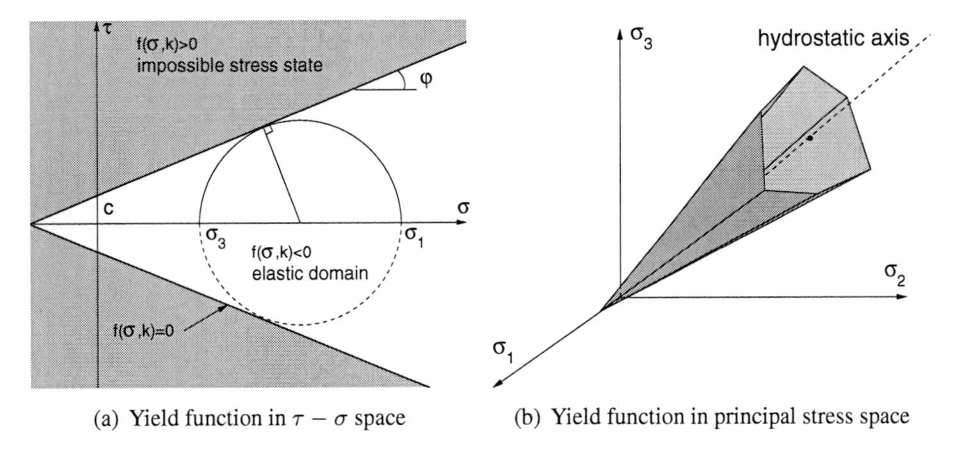

(a) Yield function in $\tau - \sigma$ space (b) Yield function in principal stress space

Figure 4.4: Illustration of the Mohr-Coulomb model

Definition of loading and unloading If the stress state fulfils the yield condition, i.e. $f(\boldsymbol{\sigma}, \boldsymbol{k}) = 0$, one must distinguish between loading and unloading. In the elasto-plastic framework the distinction between the two is accomplished by the following conditions:

For $f = 0$, loading is defined by

$$\frac{\partial f}{\partial \sigma_{ij}} d\sigma_{ij} > 0 \quad ,$$

unloading by

$$\frac{\partial f}{\partial \sigma_{ij}} d\sigma_{ij} < 0 \quad .$$

This distinction makes it possible to define different stiffnesses for both cases.[9]

Plastic potential In contrast to elastic behaviour with a functional relation between stress and strain increments, the incremental plastic strains $\Delta\varepsilon_i^{pl}$ for a plastic material at yield cannot be derived from the stress increment alone.

To define the direction of the plastic strain increment vector at a given stress state the plastic potential g is introduced; it is as function of stresses $\boldsymbol{\sigma}$ and state parameters $\boldsymbol{k}$

$$g(\boldsymbol{\sigma}, \boldsymbol{k}) = 0 \quad . \tag{4.2}$$

The state parameters $\boldsymbol{k}$ essentially control the size of the plastic potential surface and makes sure that it always passes through the current stress point.

[9] $\frac{\partial f}{\partial \sigma_{ij}} d\sigma_{ij} = 0$ is considered as neutral loading; in this case the stress increment vector lies within the yield surface.

The plastic potential helps to define a *flow rule*

$$\Delta\varepsilon_i^{pl} = \lambda \frac{\partial g(\boldsymbol{\sigma}, \boldsymbol{k})}{\partial \sigma_i} \quad . \tag{4.3}$$

The magnitude of the plastic strain increment is controlled by a scalar multiplier λ, which can be obtained from the *consistency* condition.

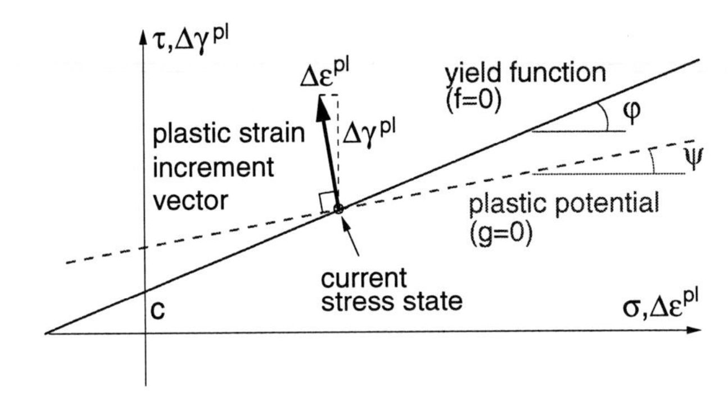

Figure 4.5: Illustration of the plastic potential g

Equation (4.2) can be plotted in the same diagram as the yield function, because coaxiality was assumed. The outward normal vector on the plastic potential surface (i.e. the plot of $g = 0$) provides relative magnitudes of the plastic strain components. As illustrated in Fig. 4.5, the plastic potential surface always passes through the current stress point; its gradient is equal to the dilation angle ψ.

The case $f = g$ is called *associated flow*. For the Mohr-Coulomb model, this implies $\varphi = \psi$. The assumption of associated flow is attractive for finite element analysis, because it leads to symmetric constitutive and global stiffness matrices, which can be inverted with less computational effort (storage and time) than non-symmetric matrices [102].

For frictionless material, however, experiments have shown that the assumption $\varphi = \psi$ is not realistic. Moreover, the energy dissipated during an increment of plastic strain becomes zero, if $\varphi = \psi$; this does not seem a realistic description of soil behaviour.

Material parameters For the FE calculations, described in the following, the linear elastic, perfectly plastic Mohr-Coulomb model (cf. Fig. 4.3 b) was used. It requires five parameters:

- Young's modulus E and Poisson's ratio ν describe the material behaviour in the elastic domain.

- The friction angle φ, cohesion c and dilation angle ψ govern the plastic behaviour of the material.

The elastoplastic Mohr-Coulomb model is implemented in ABAQUS.

4.2.2 Hypoplastic model

Hypoplasticity was applied to the face stability problem as more advanced constitutive model. Hypoplastic models are formulated as non-linear tensorial equations of the rate type (evolution equations), i.e. the stress rate is expressed in terms of strain rate, actual stress state and void ratio. The equations are incrementally non-linear and hold, equally, for loading and unloading. Moreover, the applied hypoplastic formulation takes pressure level and density into account.

An overview of the derivation and features of hypoplastic models was given by *Kolymbas* [69–72]. The author applied the version of *von Wolffersdorff* [138] in his calculations.

Evolution equation The applied hypoplastic formulation represents the objective stress rate[10] of the Cauchy stress $\overset{\circ}{\mathbf{T}}$ as function of the effective Cauchy stress $\mathbf{T}$, the stretching tensor $\mathbf{D}$ and void ratio e:

$$\overset{\circ}{\mathbf{T}} = \mathbf{h}(\mathbf{T}, \mathbf{D}, e) \quad . \tag{4.4}$$

In the version of *von Wolffersdorff*, (4.4) takes the form

$$\overset{\circ}{\mathbf{T}} = f_b f_e \frac{1}{\mathrm{tr}\,\hat{\mathbf{T}}^2} \left[F^2 \mathbf{D} + a^2 \hat{\mathbf{T}} \mathrm{tr}\,(\hat{\mathbf{T}}\mathbf{D}) + f_d a F (\hat{\mathbf{T}} + \hat{\mathbf{T}}^*) \sqrt{\mathrm{tr}\,\mathbf{D}^2} \right] \quad , \tag{4.5}$$

with the following definitions

[10] In brief, the *principle of material frame-indifference*, also called *objectivity*, states that stress changes $\overset{\circ}{\mathbf{T}}$ result only from deformation of the material and are *independent* of rotations of the observer or the state of reference [72].

$$\hat{\mathbf{T}} := \frac{\mathbf{T}}{\operatorname{tr}\mathbf{T}} ,$$

$$\hat{\mathbf{T}}^* := \hat{\mathbf{T}} - \frac{1}{3}\mathbf{I} ,$$

$$a := \frac{\sqrt{3}(3-\sin\varphi_c)}{2\sqrt{2}\sin\varphi_c} ,$$

$$F := \sqrt{\frac{1}{8}\tan^2\psi + \frac{2-\tan^2\psi}{2+\sqrt{2}\tan\psi\cos 3\vartheta}} - \frac{1}{2\sqrt{2}}\tan\psi ,$$

$$\tan\psi = \sqrt{3\operatorname{tr}\hat{\mathbf{T}}^{*2}} , \qquad \cos 3\vartheta = -\sqrt{6}\frac{\operatorname{tr}\hat{\mathbf{T}}^{*3}}{\left[\operatorname{tr}\hat{\mathbf{T}}^{*2}\right]^{3/2}} .$$

The factors f_d, f_e and f_b account for the stress- and density-dependence of the material behaviour. They are functions of void ratio e and mean pressure $p = -\operatorname{tr}\mathbf{T}/3$:

$$f_d := \left(\frac{e-e_d}{e_c-e_d}\right)^{\alpha} ,$$

$$f_e := \left(\frac{e_c}{e}\right)^{\beta} ,$$

$$f_b := \frac{h_s}{n}\left(\frac{e_{i0}}{e_{c0}}\right)^{\beta}\frac{1+e_i}{e_i}\left(\frac{3p}{h_s}\right)^{1-n}\left[3+a^2-a\sqrt{3}\left(\frac{e_{i0}-e_{d0}}{e_{c0}-e_{d0}}\right)^{\alpha}\right]^{-1} .$$

Material parameters The maximum, critical and minimum void ratios, e_i, e_c and e_d, obey *Bauer*'s [13] compression law:

$$\frac{e_i}{e_{i0}} = \frac{e_c}{e_{c0}} = \frac{e_d}{e_{d0}} = \exp\left[-\left(\frac{3p}{h_s}\right)^{n}\right] , \tag{4.6}$$

with void ratios at zero stress level, e_{i0}, e_{c0} and e_{d0}.

The other material parameters are the critical friction angle φ_c, granular hardness h_s and exponents α, β and n. These parameters are assumed insensitive to pressure or density.

Herle [51] illustrated how the eight material parameters can easily be determined from simple laboratory and index tests:

- The values for e_{c0} and e_{d0} are obtained from density index tests; $e_{i0} \approx 1.15 e_{c0}$.

- The critical friction angle φ_c is quantified by means of sand pile or direct shear tests. Also the sand chute tests (cf. Sec. 3.4) seem suitable.

- The granular hardness h_s and exponent n are derived via a numerical fit of (4.6) to compression test results, e.g. oedometer tests. As the fit is not unique, *Herle* put forward an empirical relation between n and grading curve characteristics (coefficient of uniformity U and mean grain size d_{50}).

- The exponent α reflects the influence of density on the peak shear strength. It is derived from the peak state of the stress-strain curve, preferably from triaxial tests.

- The parameter β, which takes stress and density dependence of the soil behaviour into account, can be derived from isotropic or oedometric compression tests.

Some typical ranges for the eight material constants are given in Tab. 4.1:

material constants	**standard test**	**common range**
φ_c	sand pile, direct shear	25° – 40°
h_s	oedometric compression	10 – 50000 MPa
n	oedometric compression	0.1 – 0.4
e_{d0}	density index test	0.2 – 1.0
e_{c0}	density index test	0.4 – 1.5
e_{i0}	related to e_{c0}	0.5 – 1.6
α	triaxial or direct shear test[11]	0.1 – 0.3
β	oedometric compression	1.0 – 4.0

Table 4.1: Standard tests to obtain hypoplastic parameters and their common range, [93]

Hypoplasticity is not implemented in the ABAQUS distribution. It can be incorporated via a user-defined subroutine (UMAT), though. In this work, a UMAT by *Fellin* and *Ostermann* [40] was used, which provides error-controlled time integration of constitutive models in rate form (cf. *Fellin* et al. [39]).

4.2.3 Model calibration

To calibrate the material models, triaxial and oedometer tests were performed in the laboratory and then simulated numerically. The laboratory tests were performed in

[11] Values taken from *Herle* [51].

common stress ranges up to 300 kPa (triaxial test) and 800 kPa (oedometer test). In addition, some index tests (minimum and maximum density, sand pile tests) served to determine suitable input parameters for both models.

The numerical study, and thus the model calibration, was *only* concerned with material $S1$.

Triaxial tests

Six triaxial tests were performed with initial relative densities $I_d \approx 0.9$ (dense). From isotropic confining pressures between 50 and 300 kPa the samples were loaded displacement-controlled up to an axial (engineering) strain of 20%.[12] Then the samples were unloaded to the confining pressure.

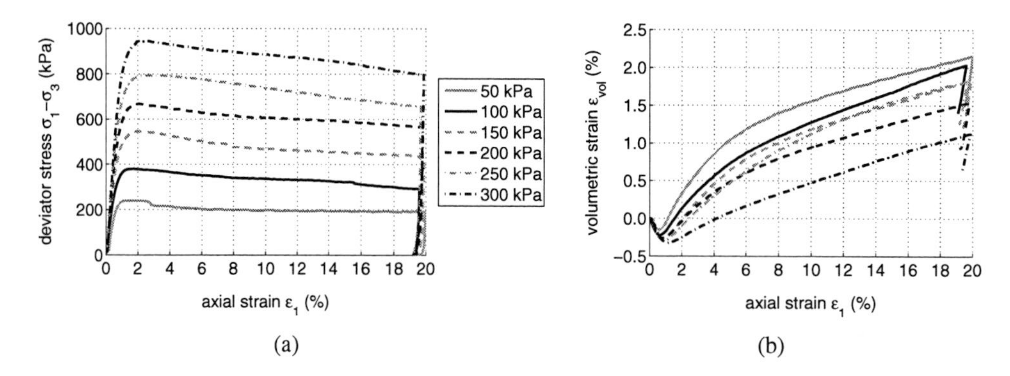

(a) (b)

Figure 4.6: Results from drained triaxial tests on dense samples ($I_d \approx 0.9$)[13]

The obtained (peak) **shear strength** parameters for the applied stress range were $\varphi_p = 36.0°$ and a cohesion $c_p = 22.9$ kPa.

It seems that none of the samples reached a critical state within the given strain range: the shear stresses still decayed with increasing axial strains (Fig. 4.6 a); also the volumetric strains had not reached constant values at the end of the tests (Fig. 4.6 b).

The **stiffness** parameter E_{load} was derived as secant modulus to the stress-strain curves for half the maximum shear stress, q_{max}. For unloading a secant modulus, E_{unload}, was obtained as indicated in Fig. 4.7. The results reveal that the stiffness grows with increasing stress level (Tab. 4.2). At a given confining pressure, E_{load} is roughly four times lower than E_{unload}.

[12]The author is well aware that the interpretation of test results for such large strains is difficult. Strictly speaking, the derived stresses and strains only have a meaning as long as the sample deforms uniformly. It is assumed that this is the case until the stress-strain curve reaches its peak.

[13]Note that the vertical axis of Fig. 4.6 was scaled up for illustration purposes.

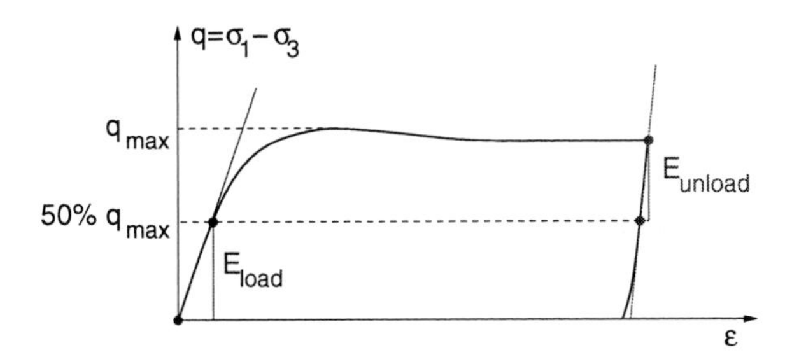

Figure 4.7: Definition of stiffness parameters for loading, E_{load}, and unloading, E_{unload}

σ_3 (kPa)	E_{load} (MPa)	E_{unload} (MPa)	b_{peak}
50	53.7	149.1	0.3706
100	55.7	198.8	0.3005
150	71.1	311.9	0.2721
200	90.1	399.0	0.2278
250	91.4	456.0	0.1939
300	104.8	459.2	0.1266

Table 4.2: Secant stiffnesses for loading and unloading and dilation b_{peak}

In a geotechnical context, the gradient of the ε_{vol} vs. ε_1 curve is often called **dilation** b with $b = \partial\varepsilon_{\text{vol}}/\partial\varepsilon_1$. The relation between b and the dilation angle ψ, in its sense as slope of the plastic potential surface, is given by

$$\sin\psi = \frac{b}{2+b} \quad . \tag{4.7}$$

Tab. 4.2 compiles the dilation b_{peak} at the axial strain where the stress-strain curves reached their peaks. It becomes obvious how b_{peak} decreases with increasing stress level.

The pronounced contractancy of the samples upon unloading is worth mentioning.

As a simple approximation to the experimental results, the dilatancy at peak strength can be correlated linearly with the mean effective stress, $p' = (\sigma'_1 + 2\sigma'_3)/3$:

$$b_{\text{peak}} \approx 0.427 - 0.484 \cdot 10^{-3}\, p' \text{ (kPa)} \quad . \tag{4.8}$$

Oedometer tests

To quantify more accurately the stiffness of the material for loading and unloading under confined conditions, eight oedometer tests were performed. The samples were

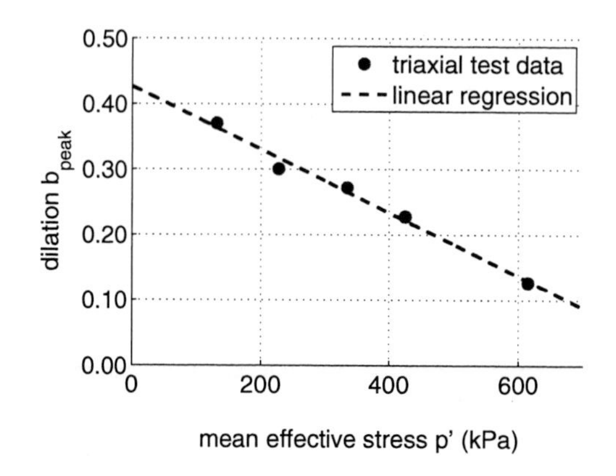

Figure 4.8: Dilatancy of the samples at different stress levels

tested with different initial densities up to vertical stresses of 785 kPa[14] (sets of four samples with $I_d \approx 0.75$ (dense) and $I_d \approx 0.16$ (loose)). Results are shown as empty circles in Fig. 4.9 together with mean values (full circles) for each set.

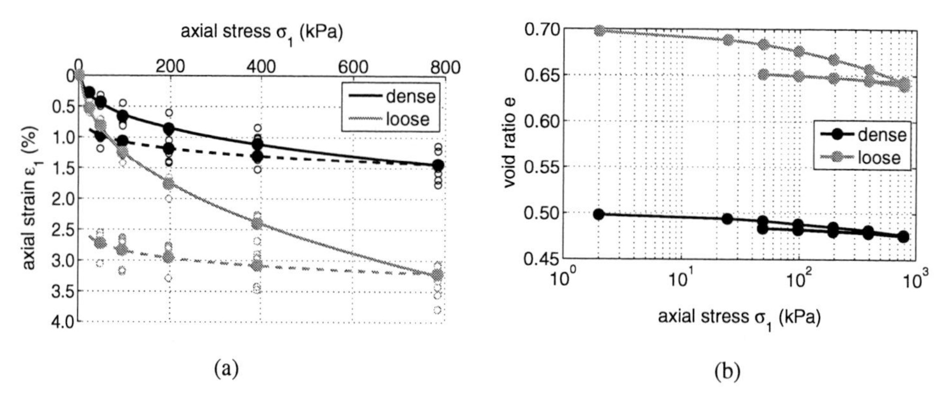

(a) (b)

Figure 4.9: Oedometer test results (mean values from four individual tests each)

To investigate the (tangent) soil stiffness, $E_{\text{oed}} = \partial\sigma_1/\partial\varepsilon_1$, at different stress levels, power law regressions were put through the ε_1 vs. σ_1 data.[15] The resulting functions were then differentiated and yielded the relations shown in Tab. 4.3 and plotted in Fig. 4.10.

The curves in Fig. 4.10 confirm that the stiffness in loading is lower than the stiffness in unloading. The difference grows with increasing stress level, and is more pronounced for the loose samples.

[14] This is, admittedly, much higher than the stress level in the sandbox experiments. Still, the obtained values serve for the model calibration.

[15] A power law for the relation between E_{oed} and σ_1 has been proposed by *Ohde* [96] or *Janbu* [57].

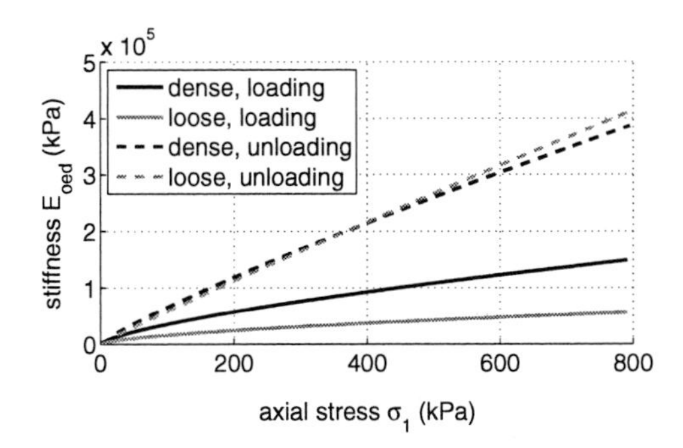

Figure 4.10: Soil stiffness as function of vertical stress

Dense samples	
loading:	E_{oed} (MPa) = 1.445 $\cdot\sigma_1(\text{kPa})^{0.694}$
unloading:	E_{oed} (MPa) = 1.262 $\cdot\sigma_1(\text{kPa})^{0.857}$
Loose samples	
loading:	E_{oed} (MPa) = 0.948 $\cdot\sigma_1(\text{kPa})^{0.612}$
unloading:	E_{oed} (MPa) = 0.774 $\cdot\sigma_1(\text{kPa})^{0.940}$

Table 4.3: Relations between soil stiffness E_{oed} and vertical stress for different initial densities in loading and unloading

Material parameters for the Mohr-Coulomb model

The parameters E and ν control the elastic behaviour of a Mohr-Coulomb material. Unfortunately, it is not easy to determine ν from standard laboratory tests. For FE calculations it has become common practice to back-calculate ν from the earth pressure coefficient at rest, K_0:

$$K_0 = \frac{\nu}{1-\nu} \quad .$$

With *Jaky*'s empirical formula,

$$K_0 \approx 1 - \sin\varphi \quad , \tag{4.9}$$

ν can be expressed as function of the friction angle:

$$\nu \approx \frac{1-\sin\varphi}{2-\sin\varphi} \quad .$$

For a friction angle φ_p = 36° above relations lead to $K_0 \approx 0.41$ and $\nu \approx 0.29$; for the loose samples with $\varphi \approx \varphi_c$ = 32°, the parameters are determined to be $K_0 \approx 0.47$ and $\nu \approx 0.32$.

In elasticity theory the stiffnesses E_{oed} and E are related via the expression

$$E = \frac{(1+\nu)(1-2\nu)}{1-\nu} E_{\text{oed}} \quad , \tag{4.10}$$

incorporating Poisson's ratio ν. With above choice of ν and K_0, and some rounding of the regression coefficients, the stiffness E (4.10) can be expressed as (non-linear) function of the stresses (Tab. 4.4).

	Dense samples (φ_p = 36°)
loading:	E (MPa) = 1.10 $\cdot\sigma_1$(kPa)$^{0.7}$ = 2.03 $\cdot\sigma_3$(kPa)$^{0.7}$
unloading:	E (MPa) = 0.96 $\cdot\sigma_1$(kPa)$^{0.9}$ = 2.05 $\cdot\sigma_3$(kPa)$^{0.9}$
	Loose samples (φ_c = 32°)
loading:	E (MPa) = 0.66 $\cdot\sigma_1$(kPa)$^{0.6}$ = 1.05 $\cdot\sigma_3$(kPa)$^{0.6}$
unloading:	E (MPa) = 0.54 $\cdot\sigma_1$(kPa)$^{0.9}$ = 1.10 $\cdot\sigma_3$(kPa)$^{0.9}$

Table 4.4: Adopted relations between soil stiffness E and stress level[17] for different initial densities in loading and unloading

These approximations are in good quantitative agreement with the values obtained from the triaxial tests (cf. Tab. 4.2), at least for the loading stage.

The dilation was approximated with (4.8),

$$\begin{aligned} b_{\text{peak}} &\approx 0.427 - 0.484 \cdot 10^{-3}\, p' \text{ (kPa)} \\ &= 0.427 - 0.484 \cdot 10^{-3}\, (2\,\sigma_3 + 1/K_0\, \sigma_3)/3 \\ &= 0.427 - 0.716 \cdot 10^{-3}\, \sigma_3 \quad \text{(for the dense samples)} \quad . \end{aligned}$$

The Mohr-Coulomb parameters for the simulation of triaxial and oedometer tests are listed in Tab. 4.5:

Test	σ_3 **(kPa)**	E **(MPa)**	ν	φ (°)	c	b_{peak}	ψ (°)
triax, dense	100	51.5	0.29	36	0	0.355	8.7
triax, dense	200	83.7	0.29	36	0	0.284	7.2
triax, dense	300	111.1	0.29	36	0	0.212	5.6
oedometer[18], loose	164	24.0	0.32	32	0	0.072	2.0

Table 4.5: Mohr-Coulomb parameters

[17] *Jaky*'s expression (4.9) was used to express σ_1 in terms of σ_3, i.e. $\sigma_1 = \sigma_3/(1-\sin\varphi)$. This makes it possible to compare the stiffnesses derived from the triaxial tests (Tab. 4.2) with those from the oedometer tests (Tab. 4.4).

[18] For the oedometer calculation, a vertical stress of 400 kPa was assumed, implying a horizontal stress of approx. 164 kPa. The dilation angle was assumed to be 2° because the loose sand in the oedometer test had a low density ($I_d \approx 0.15$).

Material parameters for the hypoplastic model

The parameter φ_c = 32° for the hypoplastic model was experimentally determined as described in Sec. 3.4. The void ratios e_{c0} = 0.75 and e_{d0} = 0.42 were determined with index tests (densest and loosest state, cf. e.g. *DIN 18126*). The maximum void ratio results from the empirical relation $e_{i0} \approx 1.15e_{c0}$ = 0.86 [51].

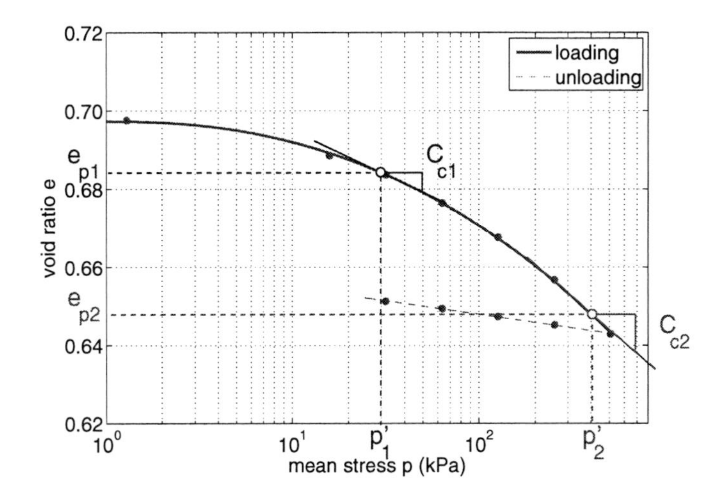

Figure 4.11: Determination of n from oedometric compression curve

Herle [51] suggested to back-calculate n from the compression curve (Fig. 4.11) via

$$n = \frac{\ln\left(\dfrac{e_{p1}}{e_{p2}}\dfrac{C_{c2}}{C_{c1}}\right)}{\ln\left(\dfrac{p_1}{p_2}\right)} , \tag{4.11}$$

which revealed $n \approx 0.30$ for stresses between 30 and 400 kPa. This n is in good quantitative agreement with values quoted by *Herle* [51]. A (numerical) fit of the compression law (4.6) to the oedometer curve (for loose sand) revealed pairs of n and h_s. For n = 0.30 a granular hardness h_s of roughly 1000 MPa was obtained.

The exponent α was determined[19] from triaxial test results of dense samples to $\alpha \approx 0.1$; β was approximated by 2.25, as suggested by *Laudahn* [76] for a similar material.

[19] See *Herle* [51] for a detailed description of the fitting procedure.

φ_c (°)	h_s (MPa)	n	e_{d0}	e_{c0}	e_{i0}	α	β
32	1000	0.30	0.42	0.75	0.86	0.10	2.25

Table 4.6: Hypoplastic parameters

4.2.4 Element test simulation

To check the performance of both material models, a triaxial test and an oedometer test were simulated numerically. A special focus was put on the behaviour of the models in unloading, because the soil in the vicinity of the tunnel face was mainly unloaded in the sandbox experiments.

Triaxial test

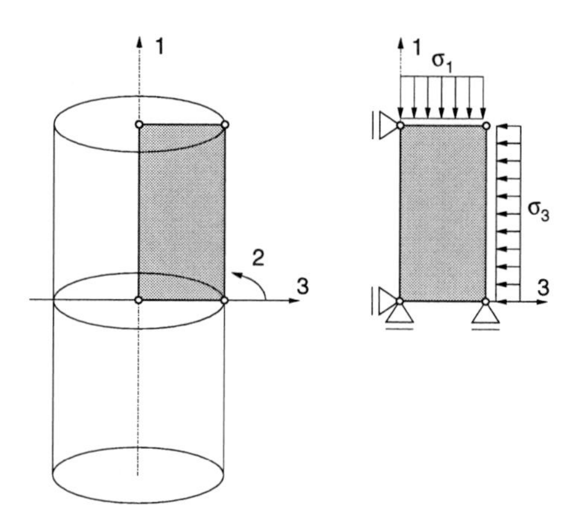

Figure 4.12: FE idealisation of a triaxial test (single element test)

The numerical model for the triaxial test (Fig. 4.12) consisted of a single axially symmetric element with 4 nodes (linear shape functions).[20] Boundary conditions were chosen as follows: the bottom nodes were fixed in vertical direction. Horizontal movements were restricted on the left boundary of the element, i.e. the symmetry axis of the sample. The lateral stress was set to $\sigma_2 = \sigma_3 = 100/200/300$ kPa. From an initially isotropic stress state, the nodes on top of the element were moved vertically downwards up to a compression of $\varepsilon_1 = 20\%$ engineering strain.[21] Then, the sample was unloaded to the initial isotropic stress.

[20] As gravity was neglected, a linear approach seemed to approximate the displacement field sufficiently.

[21] The engineering strain of $\varepsilon_1 = \Delta h/h_0 = 20\%$ is equivalent to a logarithmic strain of $\epsilon_1 = \ln((h_0 - \Delta\, h)/h_0 = 22.3\%$

Fig. 4.13 shows numerical and experimental results for a triaxial test with a confining pressure of σ_3 = 200 kPa.[22]

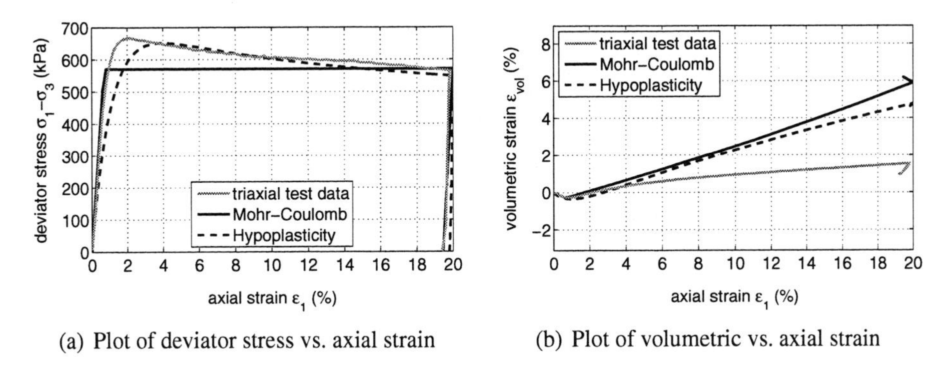

(a) Plot of deviator stress vs. axial strain (b) Plot of volumetric vs. axial strain

Figure 4.13: Comparison of numerical and experimental results for a triaxial test with σ_3 = 200 kPa

The characteristic of the experimentally observed stress-strain behaviour (Fig. 4.13 a) is not reproduced well by the Mohr-Coulomb model. It is not able to reproduce the peak of the curve and the subsequent softening, because no hardening/softening law was taken into account. The hypoplastic model, in contrast, is able to capture the strain softening behaviour of the sand. The predicted peak strength compares well to the experimental results.

The initial stiffness is captured better by the Mohr-Coulomb model; hypoplasticity predicts a behaviour that is slightly too soft. But the latter predicts better the higher stiffness for unloading.

The dilatancy of the sand (cf. Fig. 4.13 b) is overestimated by both models to roughly the same amount. The contractancy upon unloading is only predicted by the hypoplastic material law, though; the Mohr-Coulomb model shows further dilation in the unloading stage.

Oedometer test

The oedometer is characterised by the lateral confinement of the sample. Therefore, radial movements of the applied axially symmetric finite element were prohibited (Fig. 4.14). In agreement with the load stages applied in the laboratory, the top of the sample was loaded vertically to a stress of 785 kPa, then unloaded to 50 kPa and reloaded to 785 kPa.

The numerical and experimental results for the oedometer test (Fig. 4.15) confirm that the Mohr-Coulomb model is stiffer at the beginning of the loading stage than the hypoplastic model. The slope of the ε_1 vs. σ_1 curve for the Mohr-Coulomb model

[22] Please note that *engineering* strains are plotted.

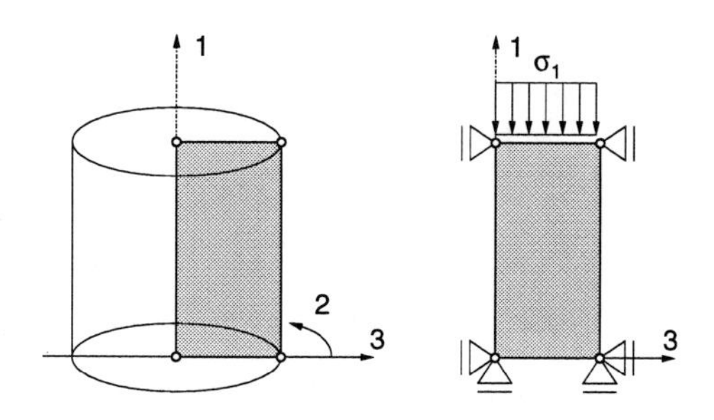

Figure 4.14: FE idealisation of an oedometer test (single element test)

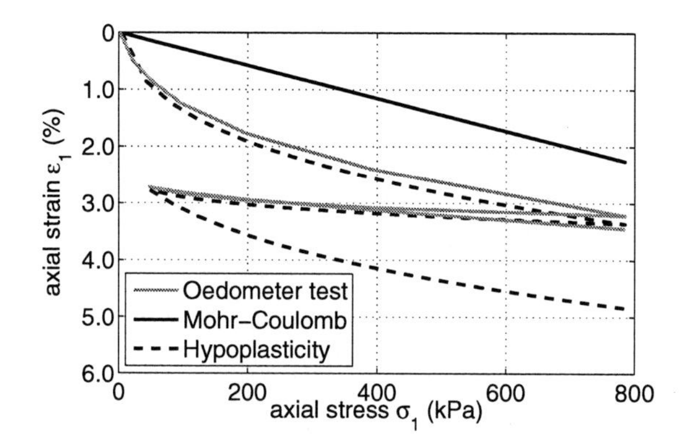

Figure 4.15: Comparison of numerical and experimental results for the oedometer test

coincides with the measured one for the calibrated stress of approx. 400 kPa. The curve follows an identical path for loading, unloading and reloading which clearly does not reflect the observed soil behaviour.

The hypoplastic model, in contrast, captures much better the non-linear stress-strain behaviour with different stiffnesses for loading and unloading. The reloading behaviour is not predicted well, though: the stiffness is too small. It must be admitted that the quantitative agreement is so good, because the material parameters were fitted to the shown experimental data.

Evaluation

The Mohr-Coulomb model has its main weakness in the realistic prediction of displacements. Various authors have shown, though, that it is applicable to limit load calculations for which displacements play a subordinate role (e.g. *Vermeer* and *Van Langen* [133] or *Zienkiewicz* et al. [140]). This is reflected in the reasonable prediction of the peak strength for the triaxial test.

Another disadvantage of the Mohr-Coulomb model is that the input parameters have to be adapted to the applied stress level and density of the sample; this task is not straightforward, if the behaviour of soil and structure are not known in advance.

Hypoplasticity predicts the different stiffnesses for loading and unloading in the oedometer test simulation, and the strain softening as well as the volumetric behaviour in the triaxial test much better. Because hypoplasticity has void ratio e as state parameter, the influence of stress level and density is already incorporated. *One* set of parameters is characteristic for a material under a variety of loading scenarios.

4.3 Parametric study for the numerical sandbox model

The simulation of the sandbox tests included a preliminary parametric study for the numerical model: as a first step, three different modelling procedures for the given problem were tested with both, the Mohr-Coulomb and the hypoplastic model. Then, the numerical configuration of the favoured procedure, e.g. time and space discretisation, element type and integration schemes, was varied. Finally, a mesh study served to assess the influence of the finite element mesh on the results of the simulations.

General setup of the numerical model

Throughout the parametric study, input parameters for both material models and boundary conditions were kept constant.

The "loose" sandbox experiments served as reference for the numerical simulations. As they were performed with densities $I_d = 0.27 \ldots 0.33$ (corresponding to void ratios e between 0.64 and 0.66), the self-weight of the soil was set to $\gamma = 16.0$ kN/m^3 ($e_0 = 0.65$, $I_d = 0.3$).

According to Sec. 3.4 the friction angle for the Mohr-Coulomb model was expected in the interval $\varphi_c = 32° < \varphi < \varphi_p = 32°{+}14°\ I_d = 36°$. The author chose the interval middle, i.e. $\varphi = 34°$.

Thus, the earth pressure coefficient was approximated by $K_0 = 1 - \sin\varphi = 0.44$. Poisson's ratio $\nu = 0.31$ resulted from matching *Jaky*'s expression with elasticity theory, as described in the previous section. A negligible cohesion of 0.005 kPa was included to enhance numerical stability. The dilation angle ψ was set to 2°.

The stiffness E was approximated with the equation for unloading of a loose sample (cf. Tab. 4.4):

$$\begin{aligned} E(\text{MPa}) &= 0.54 \cdot \sigma_1(\text{kPa})^{0.9} \\ &= 0.54 \cdot (16.0\ \text{kN/m}^3 \cdot 0.1\ \text{m})^{0.9} \\ &\approx 0.825 \quad , \end{aligned}$$

for a vertical stress σ_1 = 1.6 kPa at the crown level of the model tunnel. The parameters for the hypoplastic model were *not* adapted for the low stress level.

Tab. 4.7 summarises the applied input parameters for both models.

Mohr-Coulomb model

E (kPa)	ν	φ (°)	c (kPa)	ψ (°)
825	0.31	34.0	0.005	2

Hypoplastic model

φ_c (°)	h_s (MPa)	n	e_{d0}	e_{c0}	e_{i0}	α	β
32	1000	0.30	0.42	0.75	0.86	0.10	2.25

Table 4.7: Applied material parameters

The **spatial discretisation** of the numerical model was chosen in accordance with the performed experiments, obeying recommendations by *Ruse* [107] (Fig. 4.16 a). The system's symmetry was accounted for. For the preliminary investigation C/D was equal to 1.0.

The **time discretisation** also plays an important role for non-linear FE calculations. In simple terms, the smaller the time step, the more accurate the numerical solution. ABAQUS provides an automatic control of the time increment, which is based on tolerances for the largest residual force in an increment and the largest displacement correction. It is also possible to prescribe the number and size of the time increments. In the preliminary study automatic time incrementation was used; the maximum allowable time increment was 0.1.

For details on the numerical solution algorithm the reader is referred to the ABAQUS-manual [1].

Fig. 4.16 b illustrates the chosen **boundary conditions**: on the symmetry plane displacements in 2-direction were prohibited, same as on the back of the model. Displacements in 1-direction were restricted on the left and right planes, whereas the bottom boundary was fixed in vertical direction.

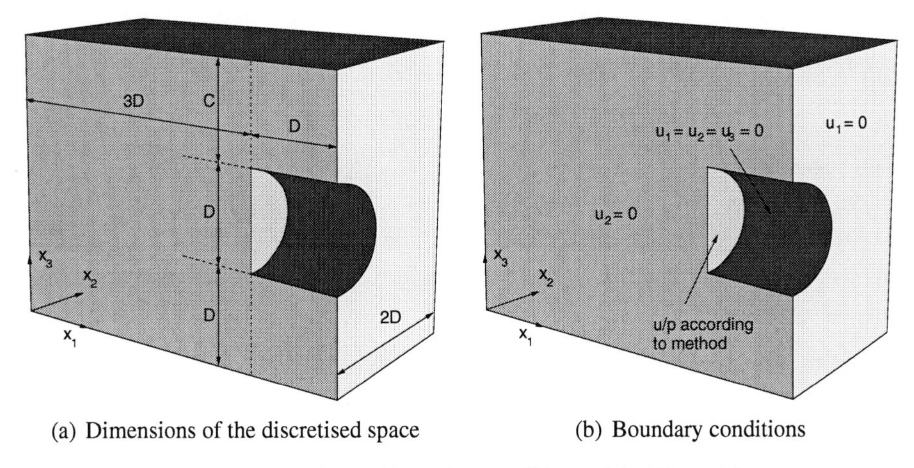

(a) Dimensions of the discretised space (b) Boundary conditions

Figure 4.16: Size and boundary conditions of the FE model

The tunnel lining was considered rigid and rough. Therefore, the nodes on the tunnel perimeter were fixed in all directions. In agreement with the experimental investigation, the construction process was not modelled, i.e. the tunnel was "wished-in-place".

The preliminary finite element mesh consisted of 670 brick elements with a total of 10461 degrees of freedom (Fig. 4.17 a). The applied elements are defined by 20 nodes, eight on the element corners and twelve in the middle of the sides (Fig. 4.17 b). Displacements are approximated with quadratic interpolation functions; for the preliminary study reduced integration was used. ABAQUS refers to the applied elements as *C3D20R* elements (*C*: continuum, *3D*: three-dimensional, *20*: 20 nodes per element, *R*: reduced integration).

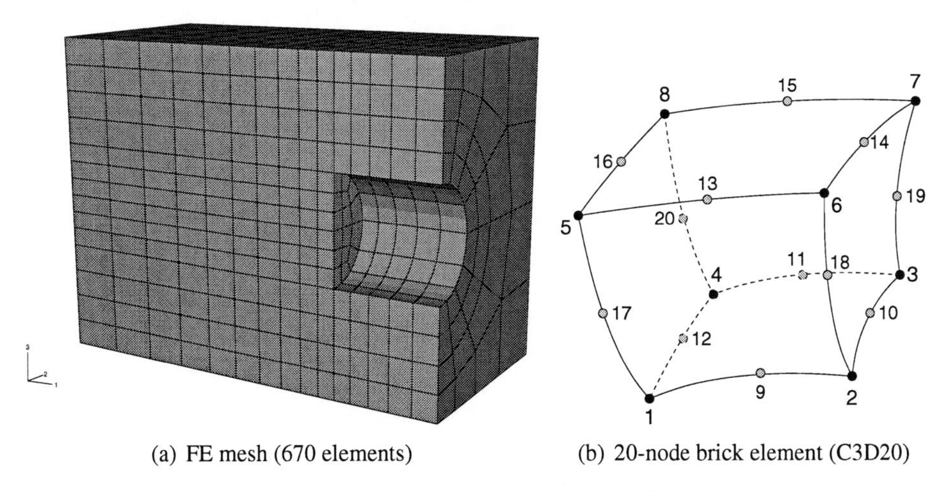

(a) FE mesh (670 elements) (b) 20-node brick element (C3D20)

Figure 4.17: FE mesh and element type for the preliminary study

The **analysis steps** depended on the modelling procedure (Sec. 4.3.1). For all model steps a large-displacement formulation was used.

Evaluation criteria

In the preliminary study the numerically predicted (dimensionless) support pressure $N_D = p_f/(\gamma\ D)$ served as main evaluation criterion for the comparison of different model configurations.

The computation times on the applied SGI Altix 350 high performance compute server (32 Intel Itanium II Processors with 1500 MHz, 128 GB Main Memory) were an additional criterion.

4.3.1 Modelling procedures for face stability analysis

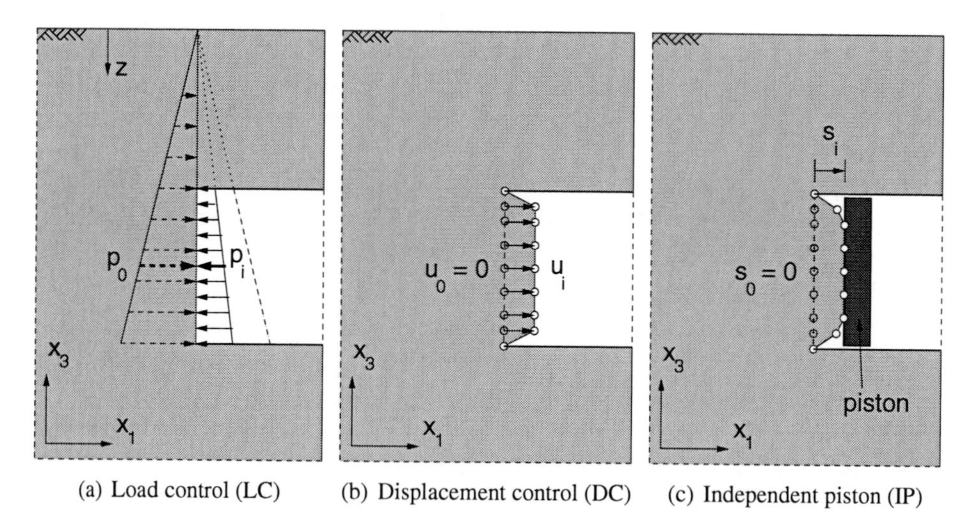

(a) Load control (LC) (b) Displacement control (DC) (c) Independent piston (IP)

Figure 4.18: Methods to trigger the face collapse numerically

The author investigated three different methods to trigger face collapse in the FE simulation:

- **Load control (LC):** A load boundary condition was applied to the tunnel face (Fig. 4.18 a). In the initial **analysis step**, this load p was in equilibrium with the horizontal stresses from the self-weight of the soil, $p = p_0$:

$$p_h(z) = K_0\ \gamma\ z \quad .$$

  In subsequent steps, p was reduced by decreasing the earth pressure coefficient K_0. In doing so, the linear increase of p with depth was preserved.

The LC model reflects most realistically the conditions in the excavation chamber of a slurry shield machine. This method was also favoured by *Ruse* [107].

- **Displacement control (DC):** In the DC model the horizontal displacements of the face nodes were prescribed (Fig. 4.18 b).

  In the initial **analysis step** the face nodes were fixed in 1-direction ($u_1 = 0$). In subsequent steps the displacement in 1-direction of these nodes was prescribed. Increments of 0.25 mm were used for each analysis step.

  As the laboratory experiments were performed displacement-controlled, the DC model reflects better the conditions in the performed sandbox experiments.

  The main disadvantage of this method is that large displacement gradients are generated in the vicinity of the tunnel contour, which can lead to numerical problems.

- **Independent piston (IP):** In a third model, the piston was modelled as independent part (Fig. 4.18 c). The piston had linear elastic properties, with a stiffness roughly five orders of magnitude higher than the soil stiffness. The diameter of the piston was slightly smaller than the tunnel diameter. Piston and soil interacted via a contact law, which allowed for separation of the elements. The contact between piston and soil was frictionless.

  In the initial **analysis step** the piston front was aligned with the front of the tunnel. Subsequently, the displacement of the piston in 1-direction was prescribed. Again, increments of 0.25 mm were used for each analysis step.

  The IP model overcomes possible numerical problems of the DC model, because soil and piston are allowed to separate. The model comes closest to the performed sandbox experiments.

Results and conclusion

The results of six finite element calculations (3 modelling procedures $\times$ 2 material models) were evaluated as **load-displacement curves** for the tunnel face.

For the LC model p was prescribed, s was taken as mean horizontal displacement of the face nodes. For the DC and the IP model s was prescribed, and the resulting support forces on the face/piston were summed up and divided by (half) the face area to calculate an equivalent support pressure p.

Fig. 4.19 shows plots of dimensionless support pressure $p/(\gamma\ D)$ vs. dimensionless face displacement s/D.

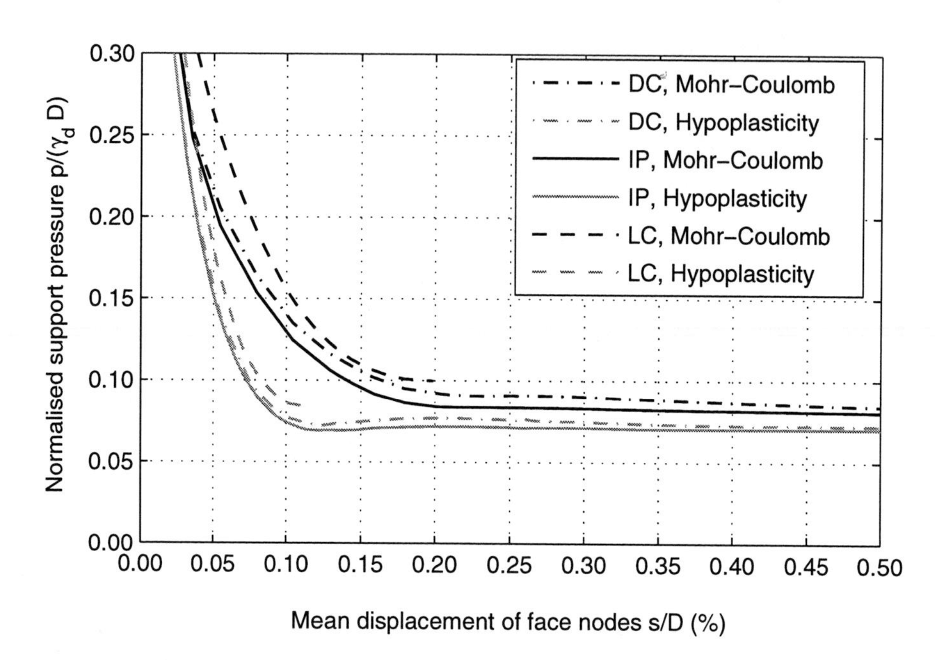

Figure 4.19: Load-displacement curves obtained from preliminary FE study

The curves for both DC analyses still show a falling trend for displacements s/D larger than 0.25...0.30%; the LC analyses stopped at mean face displacements between 0.12% (hypoplasticity) and 0.20%. Although all modelling procedures revealed the expected generic shape of the load-displacement curve, only the curves for the IP analyses reached a constant residual value in the considered range of displacements.

Computation times on the applied SGI Altix server were in the order of minutes (Tab. 4.8)[23]. Calculations with the hypoplastic model took roughly five times longer than with the Mohr-Coulomb model.

	DC	LC	IP
Mohr-Coulomb	3.1	6.1	5.2
Hypoplasticity	19.9	69.4	20.4

Table 4.8: Computation times (in min) for the different model configurations and material models

Summarising, all modelling procedures are applicable to the given problem. With reasonable computation times good qualitative predictions are achieved.

The author decided to continue the numerical investigation with the independent piston model (IP) because it comes closest to the sandbox experiments performed in the laboratory.

[23]The load-controlled calculations did not converge.

4.3.2 Influence of numerical configuration

In a second step, the numerical configuration of the piston model was investigated. In each case one "parameter" was changed with respect to the reference model:

- **Space discretisation:** The number of finite elements was increased by refining the mesh (Fig. 4.20 a). For this purpose each brick element was subdivided into smaller elements, thus increasing the number of elements from 789 to 4023 (including the piston elements).
- **Time discretisation:** The maximum time increment was reduced from 0.1 to 0.001.
- **Integration procedure:** The integration method was switched from reduced to full integration.
- **Element type:** Instead of 20-noded brick elements, 10-noded tetrahedral elements were used (Fig. 4.20 b and c). The tetrahedral elements have the advantage that mesh refinement is much easier than with brick elements.

Tab. 4.9 summarises the different configurations for the numerical investigation.

	element type	**number of elements**	**degrees of freedom**	**maximum time step**
Reference model	C3D20R	789	11823	0.1
Fine mesh	C3D20R	**4023**	55767	0.1
Full integration	**C3D20**	789	11823	0.1
Reduced time step	C3D20R	789	11823	**0.001**
Tetrahedral elements	**C3D10R**	2296	11283	0.1

Table 4.9: Computation times (in min) for the different model configurations and material models

Results and conclusion

The results of ten finite element calculations (5 model configurations $\times$ 2 material models) are plotted as load-displacement curves in Fig. 4.21.

It becomes obvious that the shape of the curves is the same for each material model. But the value for the residual pressure on the piston depends on the model configuration. Tab. 4.10 presents a comparison of the 10 simulations in terms of computation time and deviation of the resulting $N_D = p_f/(\gamma\ D)$ from the one obtained with

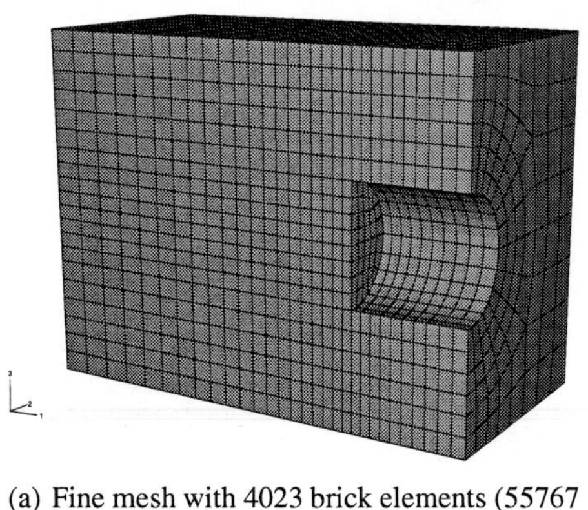

(a) Fine mesh with 4023 brick elements (55767 degrees of freedom)

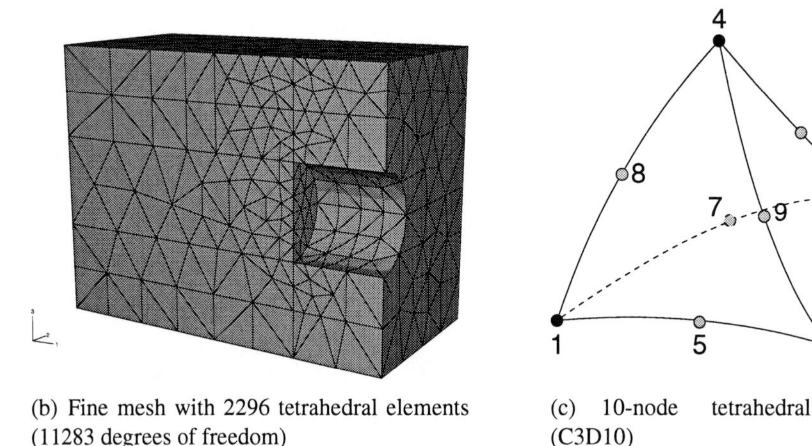

(b) Fine mesh with 2296 tetrahedral elements (11283 degrees of freedom)

(c) 10-node tetrahedral element (C3D10)

Figure 4.20: FE meshes for the model configuration

the reference model. The deviation is expressed in terms of a relative deviation = $(N_{D,\text{Variation}} - N_{D,\text{Reference}})/N_{D,\text{Reference}}$.

The results indicate that the integration scheme, the time step reduction and the element type have a minor influence on the predicted N_D values. The relative deviation of the respective changes on the result is below $\approx 6\%$. This deviation is acceptable for the purpose of the present study, as the laboratory experiments had an accuracy in the same order of magnitude.

The space discretisation with smaller elements had the major influence on the numerical predictions. Therefore, the mesh dependency of the FE results was studied in a third step of the preliminary investigation.

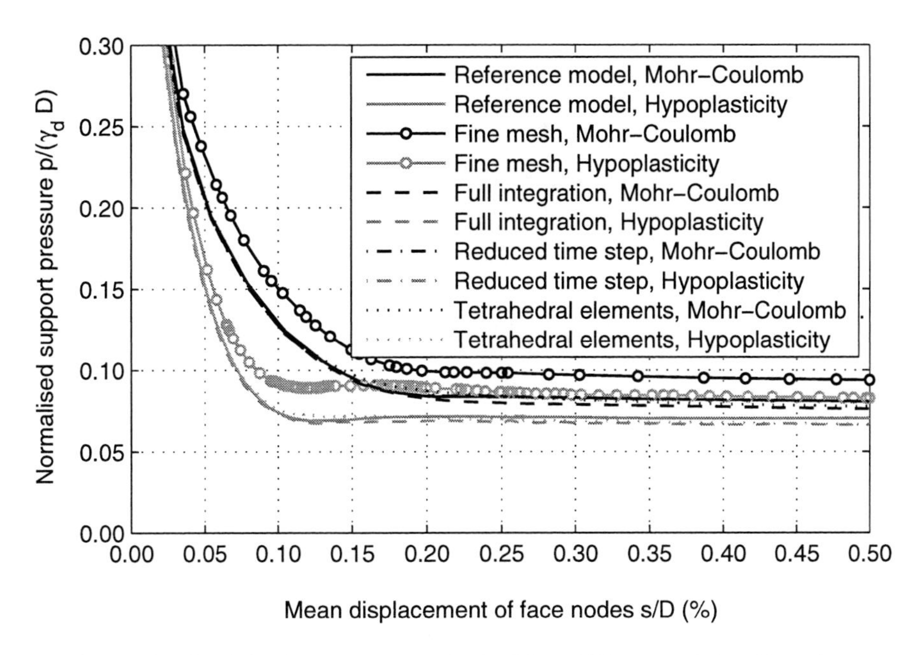

Figure 4.21: Load-displacement curves obtained from different model configurations

	computation time (min)	N_D	**rel. dev. (%)**
	Mohr-Coulomb		
Reference model	5.2	0.081	
Fine mesh	102.8	0.094	16.2
Full integration	8.2	0.076	-5.7
Reduced time step	112.1	0.081	0
Tetrahedral elements	6.2	0.078	-3.8
	Hypoplasticity		
Reference model	20.4	0.070	
Fine mesh	668.7	0.083	17.5
Full integration	43.5	0.066	-5.7
Reduced time step	153.6	0.071	0.3
Tetrahedral elements	33.1	0.067	-5.3

Table 4.10: Computation times and predicted N_D values for the different model configurations and material models

4.3.3 Influence of mesh size

As both material models showed a similar sensitivity with respect to the numerical configuration parameters, the mesh study was only performed with the Mohr-Coulomb model. This had the advantage of lower computation times, so that the mesh study could be performed with both, brick and tetrahedral elements.

The different refined meshes are shown in Fig. 4.22. The mesh refinements concentrated on areas with significant stress changes and displacements, i.e. the vicinity of the tunnel face.

Tab. 4.11 compiles further information on the investigated mesh types.

	element type	**number of elements**	**degrees of freedom**
Brick elements, mesh 1 (Reference mesh)	C3D20R	789	11823
Brick elements, mesh 2	C3D20R	1438	20307
Brick elements, mesh 3	C3D20R	2805	37869
Brick elements, mesh 4	C3D20R	3829	52395
Brick elements, mesh 5	C3D20R	6411	82011
Tetrahedral elements, mesh 1 (Reference mesh)	C3D10R	2296	11283
Tetrahedral elements, mesh 2	C3D10R	3488	18453
Tetrahedral elements, mesh 3	C3D10R	3119	16764
Tetrahedral elements, mesh 4	C3D10R	4215	21690
Tetrahedral elements, mesh 5	C3D10R	10409	51621

Table 4.11: Mesh characteristics for the mesh study

Results and conclusion

Fig. 4.23 shows the predicted load-displacement curves from the mesh study.

In terms of N_D (Tab. 4.12) there is a relative deviation of roughly 35 to 40% between the coarsest and the finest mesh. For the mesh refinement from mesh 4 to mesh 5 there is still an increase of up to 10% in the resulting N_D. The computation time increased significantly, though.

Concluding the preliminary investigation, the mesh size had the most influence on the numerical predictions of the necessary support pressure N_D. With reasonable computation times for the Mohr-Coulomb model, results with a sufficient accuracy with respect to the experimental results could be achieved.

	computation time (min)	N_D	**rel. dev. (%)**
	Brick elements (C3D20R)		
mesh 1 (ref.)	5.2	0.081	
mesh 2	9.9	0.083	2.5
mesh 3	53.7	0.096	18.3
mesh 4	150.5	0.100	24.3
mesh 5	570.5	0.109	34.3
	Tetrahedral elements (C3D10R)		
mesh 1 (ref.)	6.2	0.078	
mesh 2	20.0	0.093	19.8
mesh 3	30.5	0.095	22.5
mesh 4	34.7	0.105	35.4
mesh 5	1031.5	0.110	41.2

Table 4.12: Computation times and predicted N_D values for the mesh study with the Mohr-Coulomb model

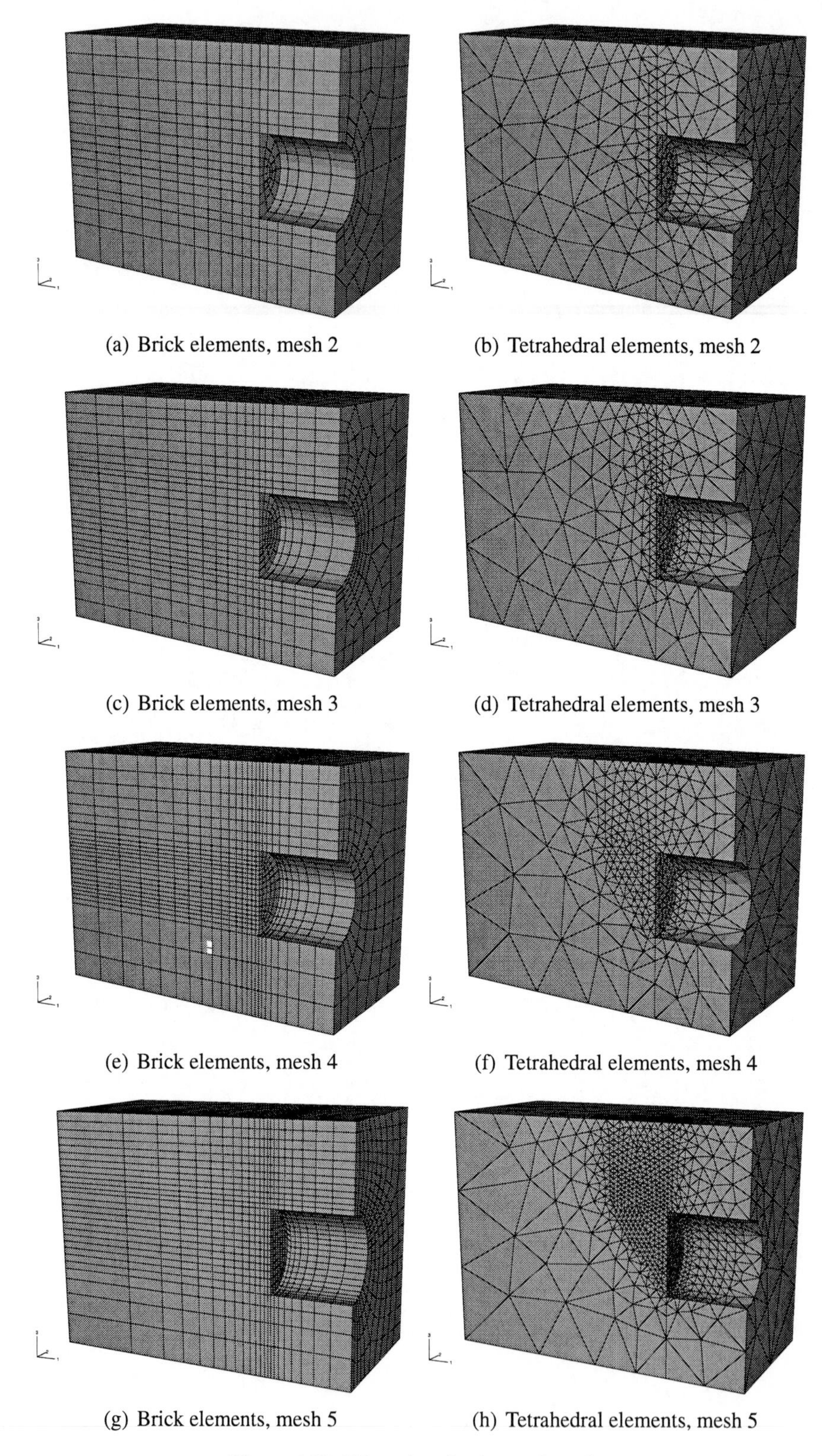

(a) Brick elements, mesh 2
(b) Tetrahedral elements, mesh 2
(c) Brick elements, mesh 3
(d) Tetrahedral elements, mesh 3
(e) Brick elements, mesh 4
(f) Tetrahedral elements, mesh 4
(g) Brick elements, mesh 5
(h) Tetrahedral elements, mesh 5

Figure 4.22: FE meshes for the mesh study

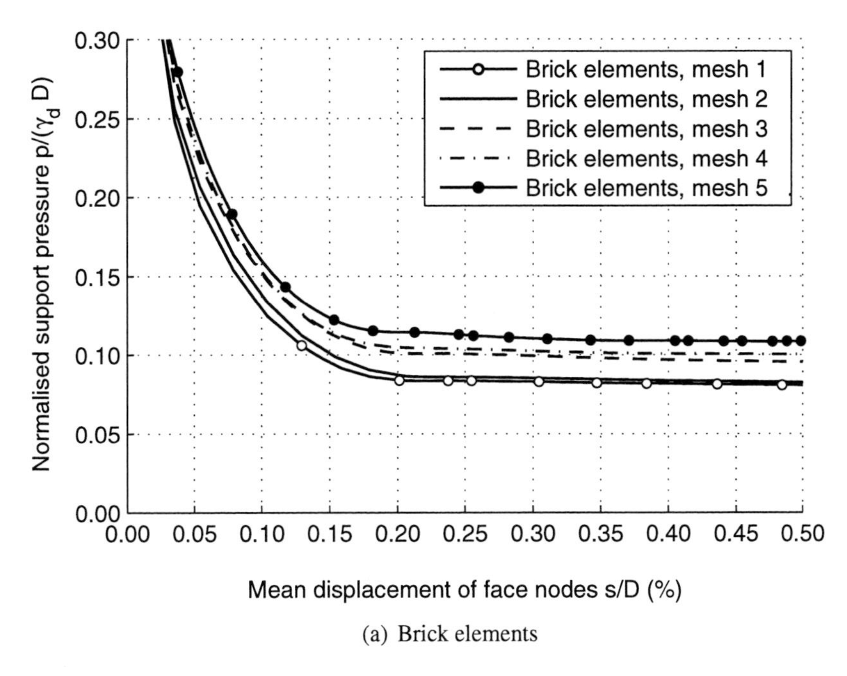

(a) Brick elements

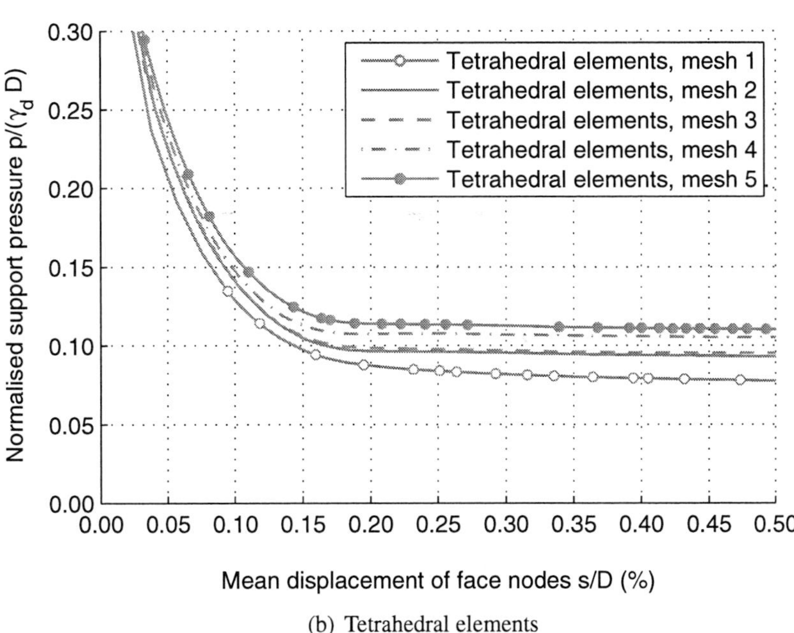

(b) Tetrahedral elements

Figure 4.23: Load-displacement curves obtained from mesh study

4.4 Finite element simulations of the sandbox tests

As a final step of the numerical face stability investigation, the performed sandbox experiments were simulated. The author concentrated on the tests with loose sand, because shear localisation was expected to play a minor role. Thus, regularisation was avoided.

Because the computation times for the hypoplastic model were expected significantly larger than for the Mohr-Coulomb model, the author decided to perform the main numerical study with **mesh 4**. Despite the potential for an increase of N_D with a finer mesh, this compromise allowed to assess the influence of the overburden C/D on the prediction of the necessary support pressure with both material models in reasonable time. The cover-to-diameter ratio was varied between 0.5 and 1.5.

Setup of the numerical model

With respect to the preliminary investigation, only the stiffness parameter E for the Mohr-Coulomb model was adapted to the vertical stress at tunnel crown level. With the approximation for unloading of a loose sample (cf. Tab. 4.4), E was 440 kPa for $C/D = 0.5$, and $E = 1190$ kPa for $C/D = 1.5$.

E (kPa)	ν	φ (°)	c (kPa)	ψ (°)
440 / 825 / 1190	0.31	34.0	0.005	2

Table 4.13: Material parameters for the Mohr-Coulomb model

For the hypoplastic model the material parameters were left as before (Tab. 4.7). Spatial and time discretisation, boundary conditions and step definition were as described in the previous section.

Evaluation criteria

In addition to the predicted N_D, the incremental displacements in the plane of symmetry served as a criterion for the comparison of the finite element calculations with the laboratory experiments.

4.4.1 Results

The load-displacement curves for a variation of C/D and both material models are shown in Fig. 4.24 for the brick elements and in Fig. 4.25 for the tetrahedral elements.

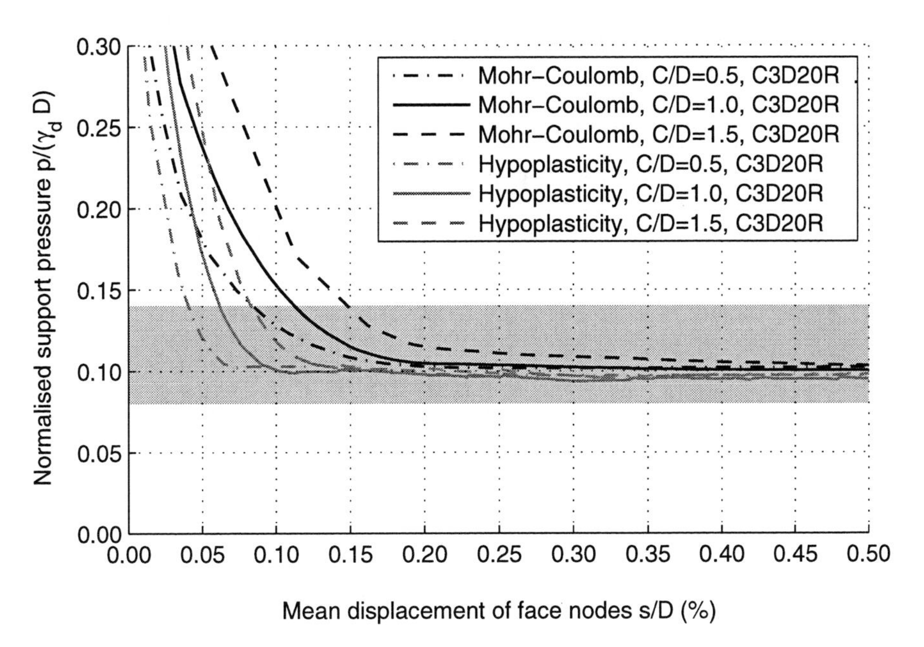

Figure 4.24: Load-displacement curves for a variation of C/D for C3D20R brick elements (range of experimental results shaded in grey)

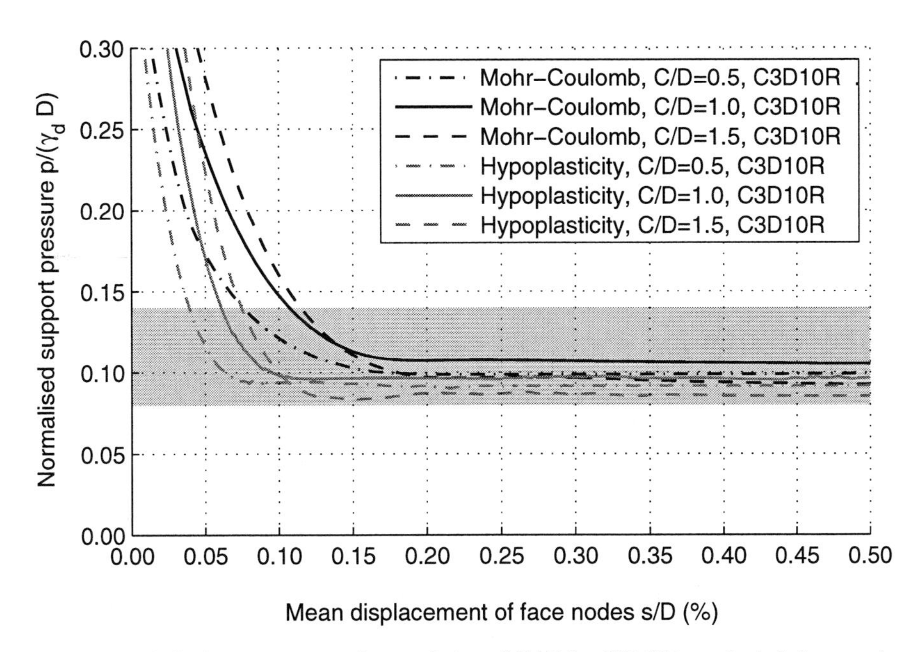

Figure 4.25: Load-displacement curves for a variation of C/D for C3D10R tetrahedral elements (range of experimental results shaded in grey)

All curves drop to approximately the same residual value of $N_D \approx 0.10$ after sufficient displacement of the piston.[24] The curves for $C/D = 1.5$ with tetrahedral

[24]The result for the Mohr-Coulomb simulation for $C/D = 1.5$ was taken at a piston advance of 1.0 mm.

elements (Fig. 4.25) seem shifted down in comparison with the other calculations. This is probably due to an inconsistent mesh refinement level.

	Mohr-Coulomb model		**Hypoplastic model**	
C3D20R	**computation time (h)**	N_D	**computation time (h)**	N_D
$C/D = 0.5$	1.18	0.103	28.14	0.098
$C/D = 1.0$	2.23	0.100	37.08	0.095
$C/D = 1.5$	92.41	0.099	118.71	0.098
C3D10R	**time (h)**		**time (h)**	
$C/D = 0.5$	0.30	0.099	5.45	0.092
$C/D = 1.0$	0.58	0.105	5.17	0.097
$C/D = 1.5$	0.58	0.092	1.34	0.085

Table 4.14: Computation times and predicted N_D values for the sandbox simulations

For the chosen input parameters and model configuration, the overburden C/D, the element type and the material model only have a marginal influence on the resulting N_D (Tab. 4.14).

Also plots of incremental displacement for an advance step from 0.25 to 0.50 mm (Fig. 4.26) reveal only a small difference between the Mohr-Coulomb and the hypoplastic model: admittedly, the magnitude of incremental displacements in the failure zone is smaller for the Mohr-Coulomb calculation. But, both models predict soil movements up to the ground surface.

4.4.2 Interpretation of results

The obtained load-displacement curves for the Mohr-Coulomb calculations are in good qualitative agreement with results published by *Ruse* [107]. In both investigations, no influence of C/D on the necessary support pressure was observed.

The absolute value for N_D is slightly smaller than predicted by *Ruse*'s empirical formula: for $\varphi = 34°$, $N_{D,\text{Ruse}} = 0.1147$. Reason for this might be that *Ruse* triggered the face-collapse load-controlled. As shown in Sec. 4.3.1, the load-controlled simulations predict a failure load which is larger than that with independent piston. The kinematics of the problem are slightly different, if the soil is allowed to bulge into the tunnel.

Sterpi and *Cividini* [116] modelled the problem with a strain softening material model. They found that neglecting strain softening, as with the Mohr-Coulomb model, led to an underestimation of displacements. This statement is in agreement with the obtained patterns of incremental displacements for the two applied models (Fig. 4.26).

For the simulations with hypoplasticity there are no published references. But the coincidence between predictions for N_D with both material models is remarkable.

Influence of dilatancy

To assess the influence of the dilation angle on the results, additional calculations with ψ = 10° and ψ = 18° were performed with the Mohr-Coulomb model for $C/D = 1.0$.

The resulting load-displacement curves (Fig. 4.27) reveal a dependency between N_D and ψ: for an increase of ψ from ψ = 2° to ψ = 18° the corresponding N_D drops from 0.100 to 0.087. This is not in agreement with findings by *Ruse* [107] who did not observe a relation between necessary support pressure and dilation angle. *Mayer* et al. [86], on the contrary, found an increase of the safety of a tunnel face for an increase of dilation angle.

The plots of incremental displacements underline the influence of ψ (Fig. 4.28): the shape of the failure zone becomes much smaller with increasing dilation angle. In all cases a tendency of the failure zone to propagate to the ground surface can be observed. Clearly, this will be the zone where the stresses fulfil the yield condition and plastic straining of the material sets on.

In the given boundary value problem the material in the failure zone has no possibility to dilate because it is laterally confined. Therefore, the stresses will increase depending on the magnitude of dilation. This stress increase leads to arching. As the dilation angle governs the increment of plastic volumetric strains, it is directly related to the arching potential of the material. This explains not only why the failure zone is smaller for ψ = 18°, but also why the resulting pressure on the piston is smaller.

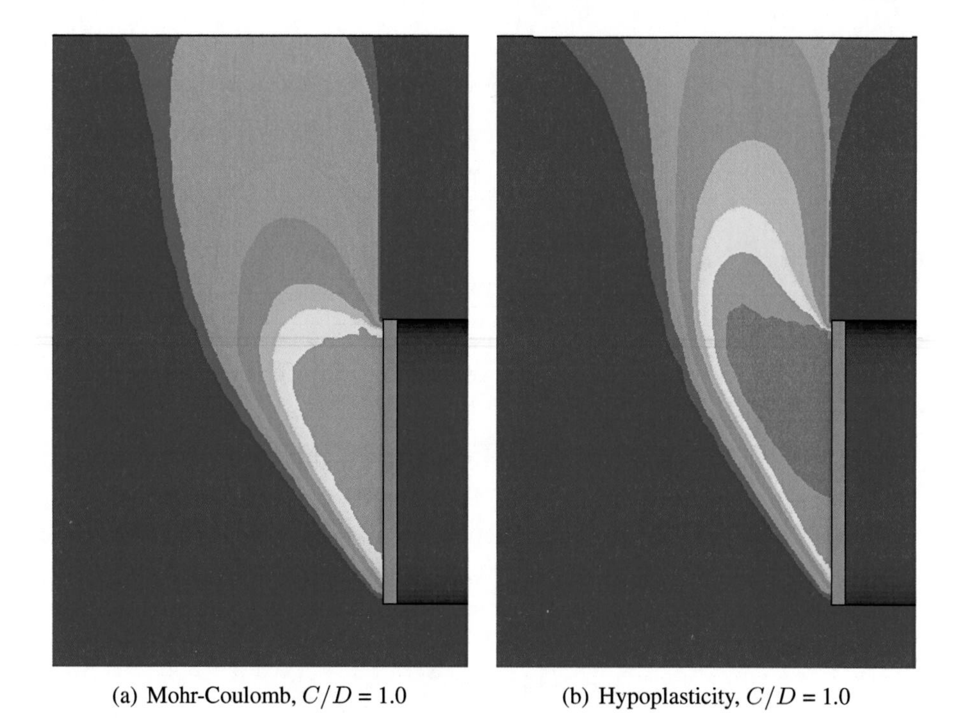

(a) Mohr-Coulomb, $C/D = 1.0$ (b) Hypoplasticity, $C/D = 1.0$

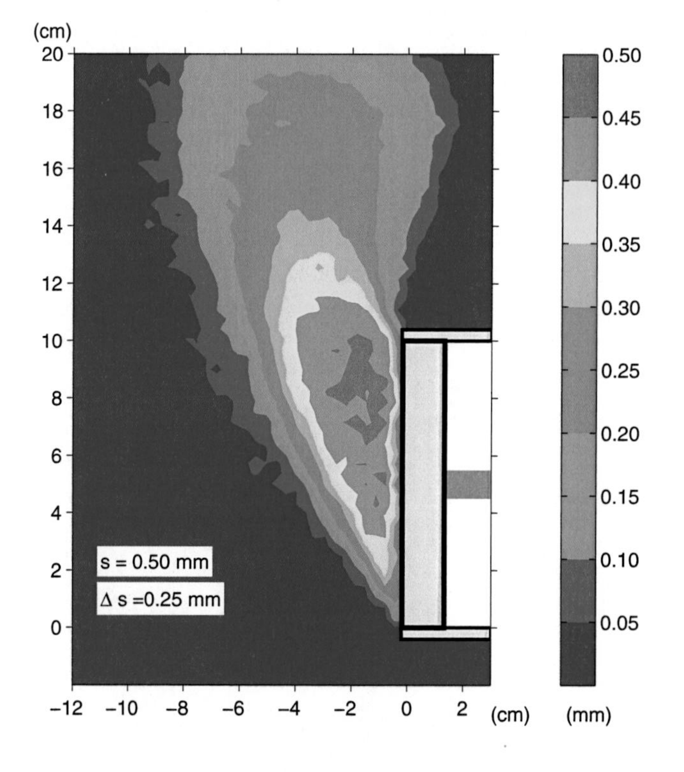

(c) PIV measurements, $C/D = 1.0$, loose material $S1$

Figure 4.26: Plots of incremental displacements for a piston advance from 0.25 to 0.50 mm

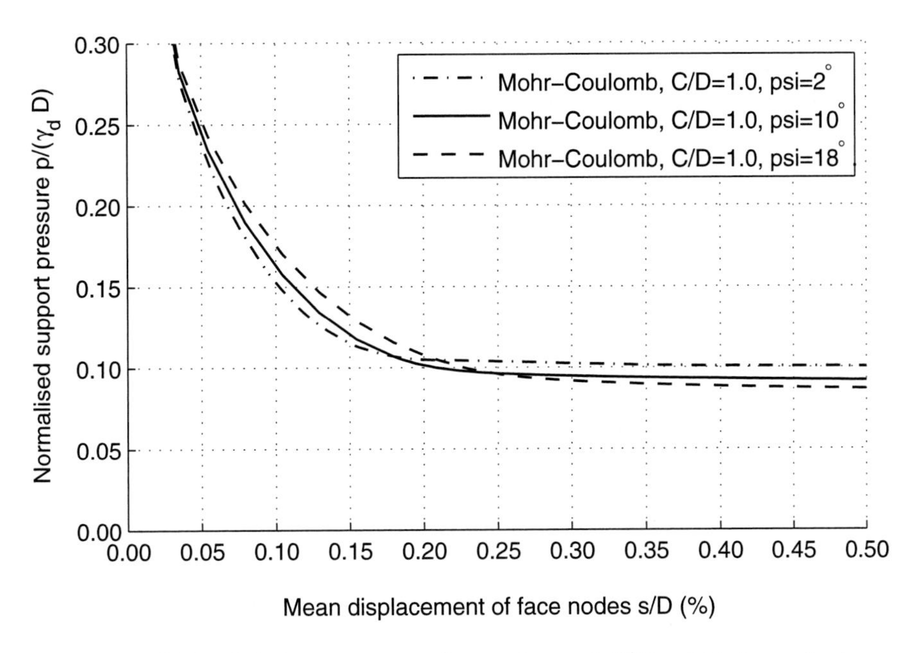

Figure 4.27: Load-displacement curves for a variation of C/D and dilation angle ψ

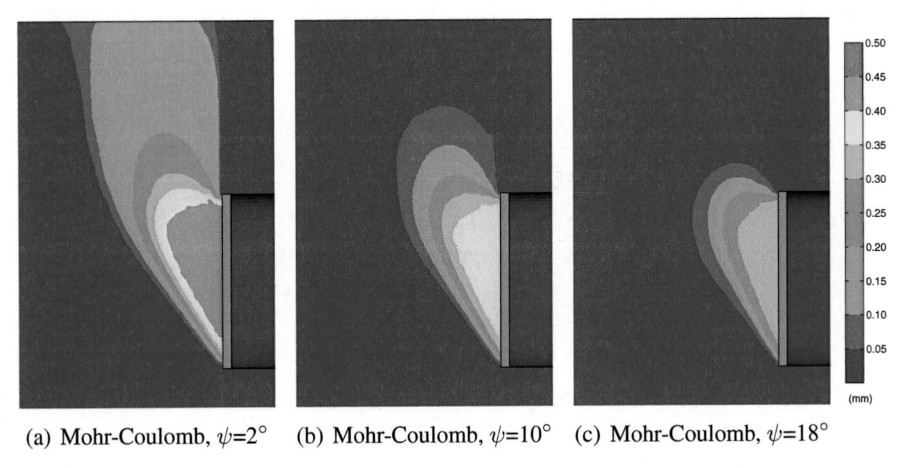

(a) Mohr-Coulomb, ψ=2° (b) Mohr-Coulomb, ψ=10° (c) Mohr-Coulomb, ψ=18°

Figure 4.28: Plots of incremental displacements for a piston advance from 0.25 to 0.50 mm for a variation of dilation angle ψ

4.5 Comparison between numerical and experimental results

The results of both the Mohr-Coulomb and the hypoplastic model are in good quantitative agreement with the measured support pressures (shaded in grey in Figs. 4.24 and 4.25).

The numerically obtained $N_D \approx 0.10$ value is roughly 10% smaller than the mean $N_D \approx 0.11$ from the laboratory experiments. A reason for this might be that mesh 4 is not fine enough: the calculation with mesh 5 revealed $N_D = 0.109$ for the C3D20R elements and $N_D = 0.110$ for the C3D10R elements (Tab. 4.12) which is in perfect agreement with the mean experimental results. As mentioned above, the author performed the final calculations with mesh 4 to compare the performance of both models in reasonable computation times.

The incremental displacements, which were observed in the PIV investigation, are predicted well by both constitutive models.

Concluding, both models are capable of predicting the necessary support pressure and the resulting displacement pattern. The Mohr-Coulomb model might seem easier to grasp, but the calibration procedure is error-prone: the expected loading history of the soil, its density and the stress level need to be considered correctly.

The hypoplastic model, in contrast, has the advantage that a single set of input parameters is sufficient for one type of soil. The effects of density and stress level on the strength of the material are incorporated in the model by means of the state parameter e. Of course, there is a price for this capacity: the finite element calculations with hypoplasticity lasted about 25 times longer than the Mohr-Coulomb calculations.

SYNOPSIS

A three-dimensional finite element investigation of face stability served to assess the ability of two constitutive models to predict the necessary support pressure *and* the displacements at collapse of the tunnel face. For this purpose, an elastoplastic Mohr-Coulomb model and a hypoplastic model were used.

For the calibration of the two models triaxial and oedometer tests were simulated. The parameters for the Mohr-Coulomb model were carefully adapted to the expected loading history of the soil, its density and the stress level. The hypoplastic model only required a single set of input parameters.

After a preliminary study into the setup of the numerical model, the small scale experiments with loose sand were simulated, varying the overburden. The observed necessary support pressure and incremental displacements were predicted sufficiently well by both constitutive models.

Chapter 5

Summary and conclusions

Summary

For tunnels driven with slurry or EPB shields the necessary support pressure in the excavation chamber must counteract water and earth pressure to prevent excessive ground movements. While the water pressure can be determined with sufficient accuracy, the determination of earth pressure at the tunnel face is rather vague. This becomes manifest in relatively high partial safety factors on the earth pressure and, also, the large number of proposed calculation methods for tunnel face stability analysis.

To give an overview, the author compiled a variety of analytical, experimental and numerical approaches on the topic. A simple calculation with some representative approaches has shown that the predicted necessary support pressures differ by as much as one order of magnitude. Also, the models' sensitivities with respect to the input parameters was significantly different. The disclosed inconsistencies raised the quest for the true collapse load for a tunnel face in cohesionless soil.

To assess the quality of the proposed approaches, the collapse of a tunnel face was modelled with small-scale model tests at single gravity. Two individual test series served to investigate the evolution of the failure mechanism and the development of the necessary support force at the face in dry sand. In both series the overburden above the model tunnel and the initial density of the soil were varied. The collapse was triggered by incrementally retracting a support piston into the model tunnel.

The investigation of the failure mechanism was performed by means of Particle Image Velocimetry, which allowed to trace particle movements throughout a test. The resulting *displacement* patterns show that the overburden has a negligible influence on the extent and evolution of the failure zone. The latter is significantly influenced, though, by the initial density of the sand: in dense sand a chimney-wedge-type collapse mechanism developed, which propagated towards the soil surface. Initially loose sand did not show any development of a discrete collapse mechanism.

The support *force* was monitored during piston displacement. For displacements larger than 1% of the tunnel diameter, all curves reached approximately the same residual value, which was, in contrast to the failure zone, neither influenced by the overburden nor the initial density of the sand. Some analytical approaches from the literature were compared to the experimental results: only the results by *Ruse* [107] and *Léca/Dormieux* [77] were able to predict the experimental results on a 95% confidence level. The others significantly overestimated the necessary support force.

In addition, a three-dimensional finite element investigation of face stability served to assess the ability of two constitutive models to predict the necessary support pressure *and* the displacements at collapse of the tunnel face. For this purpose, an elasto-plastic Mohr-Coulomb model and a hypoplastic model were used. After a preliminary study into the setup of the numerical model, the small scale experiments with loose sand were simulated, only varying the overburden. The observed necessary support pressure and incremental displacements were predicted sufficiently well by both constitutive models.

Conclusions

When starting this research work, the author had the vision of "inventing" a better theoretical model for the face stability problem. Throughout the investigation, he became aware, though, that there is a sufficient number of different models for the given problem. So the task was redefined to *evaluate* the proposed approaches on the basis of example calculations, sensitivity analysis and his own experimental and numerical investigation. The following conclusions hold for tunnels with low overburden in sand.

It became obvious that the investigated variants of the *Horn* model, which is frequently used in engineering practice, have some major drawbacks with respect to other models: they are relatively sensitive to the choice of the input parameters, and their "configuration" is not straightforward. Moreover, the investigated configuration significantly overestimates the necessary support pressure. This might seem "on the safe side", but close to the ground surface the risk of uplift and blow-outs increases.

The author found that *Ruse*'s model, which was derived from finite element calculations, provided a reasonably good fit to the experimentally obtained data in terms of necessary support pressure. So, if displacements are not important, *Ruse*'s estimation seems sufficient.

If displacements are important, none of the theoretical models is suitable, because they hardly ever take displacements into consideration. In this case, the design engineer should resort to three-dimensional finite element calculations. Although both investigated material models showed a good performance, the hypoplastic has some advantages over the Mohr-Coulomb model: for the calibration of the Mohr-Coulomb model all material parameters have to be adapted to the expected loading history of the soil, its density and the stress level. The hypoplastic model, in contrast, has the advantage that a single set of input parameters is sufficient for one type of soil.

In the author's opinion, the presented results constitute a useful frame of reference for practitioners who want to check "their" approach and, also, for researchers who want to validate their models.

Further research is necessary to extend the author's observations to soil with cohesion. This task is rather difficult, if not impossible, with small scale model tests at single gravity: the cohesion would have to be scaled down with the geometric scaling factor n, which is not possible with sand – an artificial analogous material would have to be found. For the same sand, only centrifuge tests could reveal meaningful results.

Also the effect of stabilisation measures, such as face anchors, was not touched. There are a few publications on the topic, but the shape and evolution of failure mechanisms with face anchors are hardly known. In this context small-scale model tests seem suited to reveal the qualitative behaviour of a reinforced ground.

Bibliography

[1] *ABAQUS Inc.* (2004), *ABAQUS Online Documentation: Version 6.5-1.*

[2] *Al Hallak R.*, *Garnier J.* and *Léca E.* (2000), Experimental study of the stability of a tunnel face reinforced by bolts, in *Geotechnical Aspects of Underground Construction in Soft Ground* (eds. O. Kusakabe, K. Fujita and Y. Miyazaki), pp. 65–68, Balkema, Rotterdam.

[3] *Anagnostou G.* and *Kovári K.* (1992), Ein Beitrag zur Statik der Ortsbrust beim Hydroschildvortrieb, in *Symposium '92, Probleme bei maschinellen Tunnelvortrieben?, Gerätehersteller und Anwender berichten.*

[4] *Anagnostou G.* and *Kovári K.* (1996), Face Stability Conditions with Earth-Pressure-Balanced Shields, *Tunnelling and Underground Space Technology*, **11**(2): pp. 165–173.

[5] *Anagnostou G.* and *Serafeimidis K.* (2007), The dimensioning of tunnel face reinforcement, in *ITA-AITES World Tunnel Congress "Underground Space - the 4th Dimension of Metropolises", Prague May 2007.*

[6] *Ashworth T.* (2007), *Experimentelle Untersuchung der Ortsbruststabilität von seichten Tunneln*, Research project, Innsbruck University.

[7] *Atkinson J.H.* and *Potts D.M.* (1977), Stability of a shallow circular tunnel in cohesionless soil, *Géotechnique*, **27**(2): pp. 203–215.

[8] *Babendererde T.* (2008), Personal communication.

[9] *Bakker K.*, *van Scheldt W.* and *Plekkenpol J.* (1996), Predictions and a monitoring scheme with respect to the boring of the Second Heinenoord Tunnel, in *Geotechnical Aspects of Underground Construction in Soft Ground* (eds. R.J. Mair and R.N. Taylor), pp. 459–464, Balkema, Rotterdam, cited in [19].

[10] *Bakker K.J.*, *de Boer F.* and *Kuiper J.C.* (1999), Extensive independent research programs on Second Heinenoord tunnel and Botlek Rail tunnel, in *Geotechnical Engineering for Transportation Infrastructure* (ed. Barends, F.B.J. et al.), pp. 1969–1978, Balkema, Rotterdam, cited in [19].

[11] *Balla A.* (1963), Rock pressure determination from shearing resistance, in *Proc. Int. Conf. Soil. Mech., Budapest*, cited in [120].

[12] *Balthaus H.* (1988), Standsicherheit der flüssigkeitsgestützten Ortsbrust bei schildvorgetriebenen Tunneln, in *Festschrift H. Duddeck*, Institut für Statik der Technischen Universität Braunschweig, pp. 477–492, Springer, Berlin.

[13] *Bauer E.* (1996), Calibration of a comprehensive hypoplastic model for granular materials, *Soils and Foundations*, **36**(1): pp. 13–26.

[14] *Been K.*, *Jefferies M.G.* and *Hachey J.* (1991), The critical state of sands, *Géotechnique*, **41**(3): pp. 365–381.

[15] *Bétourney M.C.*, *Mitri H.S.* and *Hassani F.* (1994), Chimneying disintegration failure mechanisms in hard rock mines, in *Rock Mechanic Models and Measurements – Challenges from Industry, Proc. 1st North Am. Rock Mech. Symp., Austin* (eds. P.P. Nelson and S.E. Auerbach), pp. 987–996, Balkema, Rotterdam, cited in [18].

[16] *Bliem C.* (2001), *3D Finite Element Berechnungen im Tunnelbau*, No. 4 in Advances in Geotechnical Engineering and Tunnelling, Logos, Berlin.

[17] *Bolton M.D.* (1986), The strength and dilatancy of sands, *Géotechnique*, **36**(1): pp. 65–78.

[18] *Brady B.H.G.* and *Brown E.T.* (2006), *Rock mechanics for underground mining*, 3rd ed., Springer, Dordrecht.

[19] *Broere W.* (2001), *Tunnel Face Stability & New CPT applications*, PhD thesis, Delft University.

[20] *Broms B.B.* and *Bennermark H.* (1967), Stability of clay at vertical openings, in *ASCE Journal of the Soil Mechanics and Foundations Division*, Vol. 93, pp. 71–94.

[21] *Bundesanstalt für Straßenwesen (BASt)* (2007), ZTV-ING: Zusätzliche Technische Vertragsbedingungen und Richtlinien für Ingenieurbauten, Teil 5 Tunnelbau, Verkehrsblatt, Ausgabe 2003 und Entwurf 2007.

[22] *Cabarkapa Z.* (2006), Discussion on Shear strength of cohesionless soil at low stress, *Géotechnique*, **56**(6): pp. 439–441.

[23] *Chaffois S.*, *Laréal P.*, *Monnet J.* and *Chapeau C.* (1988), Study of tunnel face in a gravel site, in *Proc. 6th Int. Conf. on Numerical Methods in Geomechanics, Innsbruck* (ed. G. Swoboda), Vol. 3, pp. 1493–1498.

[24] *Chambon P.* and *Corté J.F.* (1994), Shallow tunnels in cohesionless soil: stability of tunnel face, *ASCE Journal of Geotechnical Engineering*, **120**(7): pp. 1148–1165.

[25] *Chambon P., Corté J.F., Garnier J.* and *König D.* (1991), Face stability of shallow tunnels in granular soils, in *Centrifuge 91* (eds. H.Y. Ko. and F. McLean), pp. 99–105, Balkema, Rotterdam.

[26] *Chambon P., Couillaud A., Munch P., Schürmann A.* and *König D.* (1995), Stabilité du front de taille d'un tunnel: Étude de l'effet d'échelle, in *Geo 95*, p. 3, cited in [124].

[27] *Chen W.F.* and *Liu X.L.* (1990), *Limit analysis in soil mechanics*, Vol. 52 of *Developments in geotechnical engineering*, Elsevier, Amsterdam.

[28] *Cornforth D.H.* (1973), Prediction of drained strength of sands from relative density measurements. In: Evaluation of relative density and its role in geotechnical projects involving cohesionless soils, in *ASTM Spec. Techn. Publ. 523*, pp. 281–303, American Society for Testing and Materials, Philadelphia.

[29] *Davis E., Gunn M., Mair R.* and *Seneviratne H.* (1980), The stability of shallow tunnels and underground openings in cohesive material, *Géotechnique*, **30**(4): pp. 397–416.

[30] *De Beer E.* (1965), Influence of the Mean Normal Stress on the Shearing Strength of Sand, in *Proc. 6th Int. Conf. Soil Mech. Found. Engng., Montreal*, Vol. 1, pp. 165–169.

[31] *De Jong D.* (1964), Lower-bound collapse theorem and lack of normality of strain-rate to yield surface for soils, in *Rheology and Soil Mechanics, IUTAM, Symp., Grenoble* (eds. J. Kravtchenko and P. Sirirys), pp. 69–75, Springer, Berlin, cited in [27].

[32] *Dewoolkar M.M., Santichaianant K.* and *Ko H.Y.* (2007), Centrifuge modelling of granular soil response over active circular trapdoors, *Soils and Foundations*, **47**(5): pp. 931–945.

[33] *DIN 18126* (1996), German Standard "Bestimmung der Dichte nichtbindiger Böden bei lockerster und dichtester Lagerung", Beuth, Berlin.

[34] *DIN 4085* (1986), German Standard "Berechnung des Erddrucks", Beuth, Berlin.

[35] *DIN 4126* (1986), German Standard "Ortbeton-Schlitzwände", Beuth, Berlin.

[36] *Doro I.* (2006), *Theoretische und experimentelle Untersuchung der Stützkraft beim Schildvortrieb*, Diploma thesis, Innsbruck University and Trento University.

[37] *Drescher A.* and *Michalowski R.* (1984), Density variation in pseudo-steady plastic flow of granular media, *Géotechnique*, **34**(1): pp. 1–10.

[38] *Fannin R.J.*, *Eliadorani A.* and *Wilkinson J.M.T.* (2005), Shear strength of cohesionless soil at low stress, *Géotechnique*, **55**(6): pp. 467–478.

[39] *Fellin W.*, *Mittendorfer M.* and *Ostermann A.* (2008), Adaptive integration of constitutive rate equations, *Computers and Geotechnics*, submitted for publication.

[40] *Fellin W.* and *Ostermann A.* (2002), Consistent tangent operators for constitutive rate equations, *Int. J. Numer. Anal. Methods Geomech.*, **26**: pp. 1213–1233.

[41] *Fukushima S.* and *Tatsuoka F.* (1984), Strength and deformation characteristics of saturated sand at extremely low pressures, *Soils and Foundations*, **24**(4): pp. 30–48.

[42] *Gabener H.G.*, *Rodatz W.* and *Maybaum G.* (1999), Nachweis der Ortsbruststandsicherheit bei Querung tidebeeinflußter Gewässer, in *Vorträge zum 6. Darmstädter Geotechnik-Kolloquium*, No. 44 in Mitteilungen des Institutes und der Versuchsanstalt für Geotechnik, Darmstadt University of Technology, pp. 15–30.

[43] *Gaj F.*, *Guglielmetti V.*, *Grasso P.* and *Giacomin G.* (2003), Experience on Porto-EPB follow-up, *Tunnels & Tunnelling International*, **35**(12): pp. 15–18.

[44] *Gioda G.* and *Swoboda G.* (1999), Developments and applications of the numerical analysis of tunnels in continuous media, *Int. J. Numer. Anal. Methods Geomech.*, **23**: pp. 1393–1405.

[45] *Girmscheid G.* (2005), Tunnelbohrmaschinen - Vortriebsmethoden und Logistik, in *Betonkalender, Chap. 1.3*, pp. 119–256, Ernst & Sohn, Berlin.

[46] *Görtler H.* (1975), *Dimensionsanalyse*, Springer, Berlin.

[47] *Graf B.* (1984), *Theoretische und experimentelle Ermittlung des Vertikaldrucks auf eingebettete Bauwerke*, No. 96 in Veröffentlichungen des Institutes für Bodenmechanik und Felsmechanik der Universität Karlsruhe.

[48] *Gudehus G.* (2001), *Grundbau Taschenbuch, Teil 1: Geotechnische Grundlagen*, Chap. Stoffgesetze für Böden aus physikalischer Sicht, Ernst & Sohn, Berlin.

[49] *Harr M.E.* (1977), *Mechanics of particulate media. A probabilistic approach.*, McGraw Hill, New York, cited in [79].

[50] *Hauser C.* (2005), *Boden-Bauwerk-Interaktion bei parallel-wandigen Verbundsystemen*, No. 29 in Berichte des Lehr- und Forschungsgebiets Geotechnik der Universität Wuppertal, Shaker, Aachen.

[51] *Herle I.* (1997), *Hypoplastizität und Granulometrie einfacher Korngerüste*, No. 142 in Veröffentlichungen des Institutes für Bodenmechanik und Felsmechanik der Universität Karlsruhe.

[52] *Hochgürtel T.* (1998), *Numerische Untersuchungen zur Beurteilung der Standsicherheit der Ortsbrust beim Einsatz von Druckluft zur Wasserhaltung im schildvorgetriebenen Tunnelbau*, No. 32 in Veröffentlichungen des Instituts für Grundbau, Bodenmechanik, Felsmechanik und Verkehrswasserbau der RWTH Aachen.

[53] *Holzhäuser J.* (2000), Problematik der Standsicherheit der Ortsbrust beim TBM-Vortrieb im Betriebszustand Druckluftstützung, in *Beiträge anlässlich des 50. Geburtstages von Herrn Professor Dr.-Ing. Rolf Katzenbach*, No. 52 in Mitteilungen des Institutes und der Versuchsanstalt für Geotechnik, Darmstadt University of Technology, pp. 49–62.

[54] *Holzhäuser J.*, *Raleigh P.C.* and *Seeley T.R.* (2004), Messtechnische Überwachung des Tunnelvortriebs mit 4 EPS-TBMs beim ECIS-Projekt in Los Angeles, in *Seminar "Messen in der Geotechnik 2004", TU Braunschweig*.

[55] *Horn M.* (1961), Horizontaler Erddruck auf senkrechte Abschlussflächen von Tunneln, in *Landeskonferenz der ungarischen Tiefbauindustrie (German translation by STUVA, Düsseldorf)*.

[56] *Ishihara K.* (1993), Liquefaction and flow during earthquakes, 33rd Rankine Lecture, *Géotechnique*, **43**(3): pp. 351–415.

[57] *Janbu N.* (1963), Soil compressibility as determined by odometer and triaxial tests, in *Europ. Conf. Soil Mech. Found. Engng., Wiesbaden*, Vol. 1, pp. 19–25.

[58] *Jancsecz S.*, *Frietzsche W.*, *Breuer J.* and *Ulrichs K.* (2001), Minimierung von Senkungen beim Schildvortrieb am Beispiel der U-Bahn Düsseldorf, in *Taschenbuch für den Tunnelbau, Deutsche Gesellschaft für Geotechnik*, 25, pp. 165–214, VGE, Essen.

[59] *Jancsecz S.*, *Krause R.* and *Langmaack L.* (1999), Advantages of Soil Conditioning in Shield Tunnelling. Experiences of LRTS Izmir, in *Proc. International Congress on Challenges for the 21st Century* (ed. T.e.a. Alten), pp. 865–875, Balkema, Rotterdam.

[60] *Jancsecz S.* and *Steiner W.* (1994), Face support for a large mix-shield in heterogeneous ground conditions, in *Proc. Tunnelling '94*, pp. 531–550, Chapman & Hall, London.

[61] *Janssen H.A.* (1895), Versuche über Getreidedruck in Silozellen, *Zeitschrift des Vereins Deutscher Ingenieure*, **39**(35): pp. 1045–1049.

[62] *Kamata H.* and *Mashimo H.* (2003), Centrifuge model test of tunnel face reinforcement by bolting, *Tunnelling and Underground Space Technology*, **18**: pp. 205–212.

[63] *Kanayasu S.*, *Kubota I.* and *Shikibu N.* (1995), Stability of face during shield tunnelling – A survey on Japanese shield tunnelling, in *Underground Construction in Soft Ground* (eds. K. Fujita and O. Kusakabe), pp. 337–343, Balkema, Rotterdam.

[64] *Karstedt J.P.* (1982), *Untersuchungen zum aktiven räumlichen Erddruck im rolligen Boden bei hydrostatischer Stützung der Erdwand*, No. 10 in Veröffentlichungen des Grundbauinstitutes der Technischen Universität Berlin.

[65] *Katzenbach R.* and *Strüber S.* (2003), Schwierige Tunnelvortriebe im Lockerund Festgestein – Anforderungen an Erkundung, Planung und Ausführung, *Geotechnik*, **26**(4): pp. 224–229.

[66] *Kegelmann S.* (2001), *Ortsbruststabilität beim Vortrieb mit Hydroschilden*, Diploma thesis, Institut für konstruktiven Ingenieurbau, Lehrstuhl für Bauverfahrenstechnik, Tunnelbau und Baubetrieb der Universität Bochum.

[67] *Kimura T.* and *Mair R.J.* (1981), Centrifugal testing of model tunnels in soft soil, in *Proc. 10th Int. Conf. Soil Mech. Found. Engng., Stockholm*, Vol. 1, pp. 319–322.

[68] *Kirsch A.* and *Kolymbas D.* (2005), Theoretische Untersuchung zur Ortsbruststabilität, *Bautechnik*, **82**(7): pp. 449–456.

[69] *Kolymbas D.* (1977), A rate-dependent constitutive equation for soils, *Mech. Res. Comm.*, **4**: pp. 367–372.

[70] *Kolymbas D.* (1991), Computer-aided design of constitutive laws, *Int. J. Numer. Anal. Methods Geomech.*, **15**: pp. 593–604.

[71] *Kolymbas D.* (1991), An outline of hypoplasticity, *Archive of Applied Mechanics*, **61**: pp. 143–151.

[72] *Kolymbas D.* (2000), *Introduction to hypoplasticity*, No. 1 in Advances in Geotechnical Engineering and Tunnelling, Balkema, Rotterdam.

[73] *Kolymbas D.* (2005), *Tunnelling and Tunnel Mechanics*, Springer, Berlin.

[74] *Kolymbas D.* (2007), *Geotechnik – Bodenmechanik, Grundbau und Tunnelbau*, Springer, Berlin, 2nd edn.

[75] *Krause T.* (1987), *Schildvortrieb mit flüssigkeits- und erdgestützter Ortsbrust*, No. 24 in Mitteilung des Instituts für Grundbau und Bodenmechanik der Technischen Universität Braunschweig.

[76] *Laudahn A.* (2004), An Approach to 1g Modelling in Geotechnical Engineering with Soiltron, No. 11 in Advances in Geotechnical Engineering and Tunnelling, Logos, Berlin.

[77] *Leca E.* and *Dormieux L.* (1990), Upper and lower bound solutions for the face stability of shallow circular tunnels in frictional material, *Géotechnique*, **40**(4): pp. 581–606.

[78] *Lee F.H.* (2002), The philosophy of modelling versus testing, in *Constitutive and centrifuge modelling: two extremes* (ed. S. Springman), Swets & Zeitlinger, Lisse.

[79] *Locher H.G.* (1985), Anwendung probabilistischer Methoden in der Geotechnik, in *Mitteilungen der Schweizerischen Gesellschaft für Boden- und Felsmechanik; Studientagung, 4./5. Oktober 1985, Lausanne.*

[80] *Lüesse C.H.* and *Gipperich C.* (2001), Wesertunnel: Erfahrungen mit einem oberflächennahen Hydroschild großen Durchmessers, in *Unterirdisches Bauen 2001 – Wege in die Zukunft, Proceedings of STUVA meeting 2001, München*, pp. 63–67.

[81] *Mähr M.* (2005), *Ground movements induced by shield tunnelling in non-cohesive soils*, No. 14 in Advances in Geotechnical Engineering and Tunnelling, Logos, Berlin.

[82] *Maidl B.*, *Herrenknecht M.* and *Anheuser L.* (1995), *Mechanised Shield Tunnelling*, Ernst & Sohn, Berlin.

[83] *Maidl U.* (1995), *Erweiterung der Einsatzbereiche der Erddruckschilde durch Bodenkonditionierung mit Schaum*, Technisch-wissenschaftliche Mitteilungen der Ruhr-Universität Bochum, Mitteilung No. 95-4, cited in [52].

[84] *Mang H.* and *Hofstetter G.* (2000), *Festigkeitslehre*, Springer, Wien.

[85] *Mase G.E.* (2002), *Schaum's outline of theory and problems of continuum mechanics*, McGraw-Hill, New York.

[86] *Mayer P.M.*, *Hartwig U.* and *Schwab C.* (2003), Standsicherheitsuntersuchungen der Ortsbrust mittles Bruchkörpermodell und FEM, *Bautechnik*, **80**: pp. 452–467.

[87] *Meguid M.A.*, *Saada O.*, *Nunes M.A.* and *Mattar J.* (2008), Physical modeling of tunnels in soft ground: A review, *Tunnelling and Underground Space Technology*, **23**(2): pp. 185–198.

[88] *Mélix P.* (1987), *Modellversuche und Berechnungen zur Standsicherheit oberflächennaher Tunnels*, No. 103 in Veröffentlichungen des Institutes für Bodenmechanik und Felsmechanik der Universität Karlsruhe.

[89] *Miura K.*, *Maeda K.* and *Toki S.* (1997), Method of measurement for the angle of repose of sands, *Soils and Foundations*, **32**(2): pp. 89–96.

[90] *Mohkam M.* and *Wong Y.W.* (1988), Three dimensional stability analysis of the tunnel face under fluid pressure, in *Proc. 6th Int. Conf. on Numerical Methods in Geomechanics, Innsbruck* (ed. G. Swoboda), Vol. 4, pp. 2271–2278.

[91] *Mühlhaus H.B.* (1985), Lower Bound Solutions for Circular Tunnels in Two and Three Dimensions, *Rock Mech. and Rock Eng.*, **18**: pp. 37–52.

[92] *Muir Wood D.* (2004), *Geotechnical Modelling*, Spon Press, London.

[93] *Nübel K.* (2002), *Experimental and Numerical Investigation of Shear Localisation in Granular Material*, No. 159 in Veröffentlichungen des Institutes für Bodenmechanik und Felsmechanik der Universität Karlsruhe.

[94] *Nübel K.* and *Weitbrecht V.* (2002), Visualization of Localization in Grain Skeletons with Particle Image Velocimetry, *Journal of Testing and Evaluation ASTM*, **30**(4): pp. 322–329.

[95] *Oberguggenberger M.* and *Fellin W.* (2004), The fuzziness and sensitivity of failure probabilities, in *Analyzing uncertainty in civil engineering* (ed. Fellin, W. et al.), pp. 33–49, Springer, Berlin.

[96] *Ohde J.* (1939), Zur Theorie der Druckverteilung im Baugrund, *Der Bauingenieur*, **20**: pp. 451–459.

[97] *Ono K.* and *Yamada M.* (1993), Analysis of the arching action in granular masses, *Géotechnique*, **43**(1): pp. 105–120.

[98] *Ostermann A.* (2004), Sensitivity analysis, in *Analyzing uncertainty in civil engineering* (ed. Fellin, W. et al.), pp. 101–114, Springer, Berlin.

[99] *Palmer A.C.* (1966), A limit theorem for materials with non-associated flow rules, *J. Méchanique*, **5**(2): pp. 217–222, cited in [27].

[100] *Papamichos E.*, *Vardoulakis I.* and *Heil L.K.* (2001), Overburden modeling above a compacting reservoir using a trap door apparatus, *Physics and Chemistry of the Earth (A)*, **26**(1-2): pp. 69–74.

[101] *Plekkenpol J.W.*, *van der Schrier J.S.* and *Hergarden H.J.* (2006), Shield tunnelling in saturated sand – face support pressure and soil deformations, in *Tunnelling: A Decade of Progress, GeoDelft 1995-2005* (eds. A. Bezuijen and H. van Lottum), Taylor & Francis, London.

[102] *Potts D.M.* and *Zdravković L.* (1999), *Finite element analysis in geotechnical engineering – Theory*, Vol. 1, Thomas Telford, London.

[103] *Raffel M.*, *Willert C.* and *Kompenhans J.* (1998), *Particle Image Velocimetry*, Springer, Berlin.

[104] *Rainer E.* and *Fellin W.* (2006), Druckabhängigkeit des Reibungswinkels zwischen Festkörper und Sand, *Geotechnik*, **29**(1): pp. 28–32.

[105] *Ravazzini C.* (2007), *Experimental investigation of face stability in shallow tunnels*, Diploma thesis, Innsbruck University and Ancona University.

[106] *Rinawi W.* (2004), *Particle Image Velocimetry (PIV) applied on triaxial tests*, Diploma thesis, Innsbruck University and Trento University.

[107] *Ruse N.M.* (2004), *Räumliche Betrachtung der Standsicherheit der Ortsbrust beim Tunnelvortrieb*, No. 51 in Mitteilungen des Instituts für Geotechnik der Universität Stuttgart.

[108] *Schellart W.P.* (2000), Shear test results for cohesion and friction coefficients for different granular materials: scaling implications for their usage in analogue modelling, *Tectonophysics*, **324**: pp. 1–16.

[109] *Schubert P.* and *Schweiger H.F.* (2004), Zur Standsicherheit der Ortsbrust in Lockerböden, in *Proc. ISRM regional symposium EUROCK 2004 and 53rd Geomechanics Colloquium, October 7-9, 2004, Salzburg, Austria* (ed. W. Schubert), pp. 99–104.

[110] *Schwarz J.*, *Schmidt J.*, *Maidl R.* and *Handke D.* (2006), Stützdruckberechnungen beim Hydroschildvortrieb – Stand der Technik, dargestellt am City-Tunnel Leipzig, in *Beiträge zum 5. Geotechnik-Tag in München, "Geotechnik*

beim Verkehrswegebau", No. 38 in Schriftenreihe des Lehrstuhls und Prüfamts für Grundbau, Bodenmechanik, Felsmechanik und Tunnelbau der Technischen Universität München, pp. 117–136.

[111] *Smoltzcyk U.* (1969), Earth pressure reduction in front of a tunnel shield, in *Proc. 7th Int. Conf. Soil Mech. Found. Engng., Mexiko City*, Vol. 2, pp. 473–481.

[112] *Soubra A.H.* (2000), Kinematical approach to the face stability analysis of shallow circular tunnels, in *8th Int. Symp. on Plasticity, British Columbia, Canada*, pp. 443–445.

[113] *Soubra A.H.* (2000), Three-dimensional face stability analysis of shallow circular tunnels, in *Int. Conf. on Geotechnical and Geological Engineering, Melbourne, Australia, November 19-24*, pp. 1–6.

[114] *Soubra A.H.*, *Dias D.*, *Emeriault F.* and *Kastner R.* (2008), Three-dimensional face stability analysis of circular tunnels by a kinematical approach, in *Geo-Congress 2008: Characterization, Monitoring, and Modeling of GeoSystems (GSP 179), New Orleans*, pp. 894–901.

[115] *Stallmann M.* (2005), *Verbrüche im Tunnelbau – Ursachen und Sanierung*, Diploma thesis, University of Applied Sciences Stuttgart.

[116] *Sterpi D.* and *Cividini A.* (2004), A Physical and Numerical Investigation on the Stability of Shallow Tunnels in Strain Softening Media, *Rock Mech. and Rock Engng.*, **37**(4): pp. 277–298.

[117] *Stone K.J.L.* and *Muir Wood D.* (1992), Effects of dilatancy and particle size observed in model tests on sand, *Soils and Foundations*, **32**(4): pp. 43–57.

[118] *STUVA* (2001), Erarbeitung von Kriterien für den Einsatz von maschinellen Vortriebsverfahren im Tunnelbau, Auftraggeber Bundesanstalt für Straßenwesen (BASt), Tech. rep., (Studiengesellschaft für unterirdische Verkehrsanlagen e.V.).

[119] *Sveen J.* (2004), An introduction to MatPIV 1.6.1, Eprint No. 2, ISSN 0809-4403, Department of Mathematics, Oslo University.

[120] *Széchy K.* (1969), *Tunnelbau*, Springer, Wien.

[121] *Takano D.*, *Otani J.*, *Fukushige S.* and *Natagani H.* (2006), Investigation of Interaction Behavior Between Soil and Face Bolts using X-ray CT, in *Advances in X-ray Tomography for Geomaterials* (eds. J. Desrues, G. Viggiani and P. Bèsuelle), pp. 389–395, ISTE Ltd., London.

[122] *Tatsuoka F.* (1987), Discussion, *Géotechnique*, **37**(2): pp. 219–226.

[123] *Taylor D.* (1948), *Fundamentals of soil mechanics*, John Wiley, New York, cited in [74].

[124] *Technical Commitee 2 of ISSMGE – Physical Modelling in Geotechnics* (2007), Catalogue of scaling laws and similitude questions in centrifuge modelling.

[125] *Terzaghi K.* (1925), *Erdbaumechanik auf bodenphysikalischer Grundlage*, Franz Deutschke, cited in [86].

[126] *The British Tunnelling Society in assoc. with ICE* (2005), *Closed-face tunnelling machines and ground stability: a guideline for best practice*, Thomas Telford, London.

[127] *Thewes M.* (2005), Geotechnical risks for tunnel drives with shield machines, in *Rational Tunnelling - 2nd Summerschool, Innsbruck, 2005* (eds. D. Kolymbas and A. Laudahn), Advances in Geotechnical Engineering and Tunnelling, pp. 239–268.

[128] *Vardoulakis I.*, *Graf B.* and *Gudehus G.* (1981), Trap-door problem with dry sand: a statical approach based upon model test kinematics, *Int. J. Numer. Anal. Methods Geomech.*, **5**: pp. 57–78.

[129] *Verdugo R.* and *Ishihara K.* (1996), The steady state of sandy soils, *Soils and Foundations*, **2**: pp. 81–91.

[130] *Vermeer P.A.* and *Ruse N.M.* (2001), Die Stabilität der Tunnelortsbrust in homogenem Baugrund, *Geotechnik*, **24**(3): pp. 186–193.

[131] *Vermeer P.A.*, *Ruse N.M.*, *Dong Z.* and *Härle D.* (2000), Ortsbruststabilität von Tunnelbauwerken am Beispiel des Rennsteig-Tunnels, in *Tagungsband 2. TAE Kolloquium „Bauen in Boden und Fels“*, pp. 195–202.

[132] *Vermeer P.A.*, *Ruse N.M.* and *Marcher T.* (2002), Tunnel heading stability in drained ground, *Felsbau*, **20**(6): pp. 8–18.

[133] *Vermeer P.A.* and *Van Langen H.* (1998), Soil collapse computations with finite elements, *Ingenieurarchiv*, **59**: pp. 221–236, cited in [107].

[134] *Walz B.* (2006), *Der 1g-Modellversuch in der Bodenmechanik – Verfahren und Anwendung*, No. 40 in Veröffentlichungen des Grundbauinstitutes der Technischen Universität Berlin, Vorträge zum 2. Hans Lorenz Symposium.

[135] *Walz B.* (2006), Möglichkeiten und Grenzen bodenmechanischer 1g-Modellversuche, in *Entwicklungen in der Bodenmechanik, Bodendynamik und Geotechnik* (ed. F. Rackwitz), pp. 63–78, Springer Verlag, Berlin.

[136] *White D.J.*, *Take W.A.* and *Bolton M.D.* (2001), A deformation measurement system for geotechnical testing based on digital imaging, close-range photogrammetry, and PIV image analysis, in *Proc. 15th Int. Conf. Soil Mech. Found. Engng., Istanbul*, pp. 539–542, Balkema, Rotterdam.

[137] *White D.J.*, *Take W.A.* and *Bolton M.D.* (2001), Measuring soil deformation in geotechnical models using digital images and PIV analysis, in *Proc. 10th Int. Conf. on Computer Methods and Advances in Geomechanics, Tucson, Arizona*, pp. 997–1002, Balkema, Rotterdam.

[138] *von Wolffersdorff P.A.* (1996), A hypoplastic relation for granular materials with a predefined limit state surface, *Mechanics of Cohesive-Frictional Materials*, **1**: pp. 251–271.

[139] *Wu W.* (1992), *Hypoplastizität als mathematisches Modell zum mechanischen Verhalten granularer Stoffe*, No. 129 in Veröffentlichungen des Institutes für Bodenmechanik und Felsmechanik der Universität Karlsruhe.

[140] *Zienkiewicz O.C.*, *Humpheson C.* and *Lewis R.W.* (1975), Associated and non-associated visco-plasticity in soil mechanics, *Géotechnique*, **25**(4): pp. 671–689, cited in [107].

[141] *Zienkiewicz O.C.* and *Taylor R.L.* (2000), *The finite element method*, Butterworth-Heinemann, Oxford, 5th edn.

Appendix A

List of symbols

Symbol	Meaning
b	width of the pressure cushion
b	dilaton, ratio of volumetric to axial strain increment $\Delta\varepsilon_{\text{vol}}/\Delta\varepsilon_1$
c	cohesion
c_u	undrained shear strength
d_{50}	mean grain diameter
d_s	shear band width
e_h	horizontal earth pressure ordinate
e	void ratio
e_0	initial void ratio
e_{max}, e_{min}	maximum and minimum void ratio
e_i, e_c, e_d	pressure dependent maximum, critical and minimum void ratios
e_{i0}, e_{c0} and e_{d0}	maximum, critical and minimum void ratios at zero stress level
f	yield stress
f	yield function
f_b	factor of barotropy
f_d, f_e	factors of pyknotropy
g	plastic potential function
h_s	granulate hardness
k_i	sensitivity of a function with respect to the input parameters x_i
$\boldsymbol{k}$	vector of state parameters for the yield function
l	length
$\boldsymbol{m}$	vector of state parameters for the plastic potential function
m_d	mass of the dry sand in the sandbox
m_v	coefficient of volume compressibility
n	number of measurements / data points
n	porosity
n	hypoplastic material parameter
p	measurement uncertainty
p	(tunnel/face) support pressure
p	mean pressure
p_c	necessary support pressure at the crown

p_f	necessary support pressure
p_i	necessary support pressure at the invert
p_w	water pressure ordinate
q	surface surcharge
q	deviatoric stress
r	equivalent radius
s	(sample) standard deviation
s	displacement
s/D	dimensionless piston displacement
t	time (period)
$\mathbf{u}$	(incremental) displacement vector
v, V	velocity
v	specific volume
w	width
x	stochastic variable
x_i	governing quantity of a physical problem
z	vertical coordinate
A	area
C	overburden above the tunnel
C	force due to cohesion
C/D	cover-to-diameter ratio
D	tunnel diameter
D_{disk}	diameter of the aluminium disk
E	resultant force from effective earth pressure
E	Young's modulus
E_{load}	secant modulus to the stress-strain curve in loading
E_{unload}	secant modulus to the stress-strain curve in unloading
E_{oed}	stiffness of the soil under lateral confinement
F_{collapse}	collapse load
F_{nom}	nominal load of the load cell
F_{measured}	measured force
G	force due to self-weight
I_d	density index
I_1^{ε}	first invariant of the strain tensor
I_2^{ε}	second invariant of the strain tensor
K	coefficient of earth pressure
K_0	coefficient of earth pressure at rest
$\mathbf{K}$	stiffness matrix
M	width of interrogation cell
N	length of interrogation cell

N	stability ratio, proposed by *Broms* and *Bennermark*
N_D	dimensionless support pressure at failure
N_c, N_D, N_q	dimensionless coefficients
P	unsupported length of a tunnel excavation
P	support force underneath a trapdoor
R	cross correlation coefficient
R_{max}	maximum cross correlation coefficient for a given image cell correlation
R_i	residual work at time increment t_i
Q_1,Q_v	forces due to friction
S	necessary support force
$\mathbf{T}$	effective Cauchy stress
$\mathring{\mathbf{T}}$	objective rate of the effective Cauchy stress
U	circumference
U	coefficient of uniformity
V_x	coefficient of variability
V	volume of sand in the sandbox
W	force due to water pressure
dW	incremental work
α, β and γ	orientation angles of the camera for the PIV investigation
α, β	hypoplastic material parameters
α_f	inclination of the sand chute at failure of the sand
γ	self-weight
ε_{vol}	volumetric strain
$\varepsilon(t)$	strain tensor at time t
ε_1	engineering strain in axial direction
ϵ_1	logarithmic strain in axial direction
ε	incremental strains
η	safety factor
η_c, η_i	safety factor for support pressure at tunnel crown and invert
η_E, η_W	partial safety factor for earth and water pressure
ϑ	inclination of the sliding wedge
λ	scalar multiplier
μ_x	expected value of x
μ	friction coefficient
ν	Poisson's ratio
σ	(normal) stress
σ_v, σ_z	vertical stress
σ_x	standard deviation of x
σ_{ij}	covariances between variables x_i and x_j

σ_1	major principal effective stress
σ_2	intermediate principal effective stress
σ_3	minor principal effective stress
$\boldsymbol{\sigma}(t)$	stress tensor at time t
ρ_s	specific weight of the sand grains
τ	shear stress
φ	friction angle
φ_c	critical state friction angle
φ_p	peak friction angle
ψ	dilation angle
Δs	displacement increment of the piston
$\Delta\varepsilon_i^{pl}$	incremental plastic strain
ΔP	load increment
Δt	time increment
Δu	displacement increment
Π_j	dimensionless products of different powers of x_i

Appendix B

Results of the force measurements

This appendix compiles the results of the 52 force measurements in plots of normalised support pressure $p/(\gamma_d\ D)$ vs. normalised piston displacement s/D. The cover-to-diameter ratio C/D and the relative density I_d at the start of each test are quoted in the figures. The results are presented as described in Sec. 3.3.3 (cf. Fig. 3.43 on page 100).

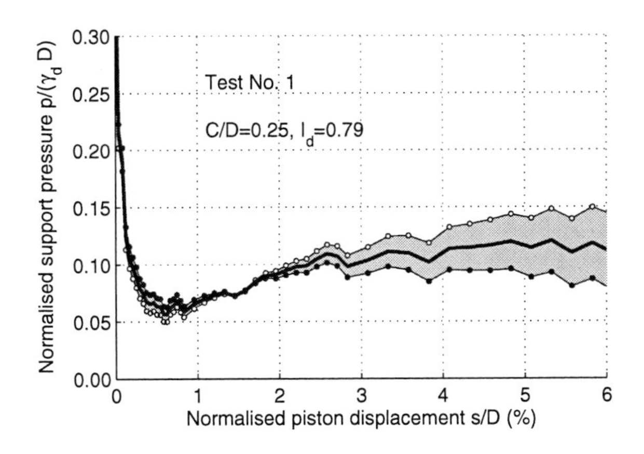

Figure B.1: Results of test No. 01

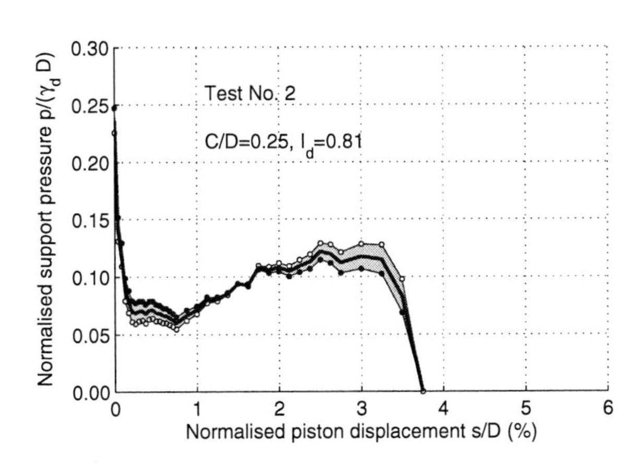

Figure B.2: Results of test No. 02

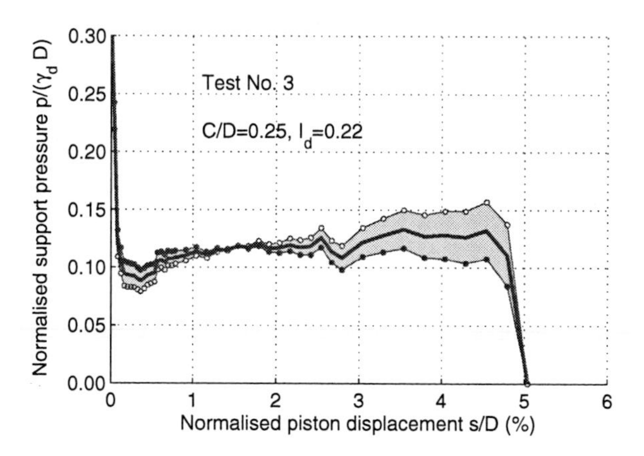

Figure B.3: Results of test No. 03

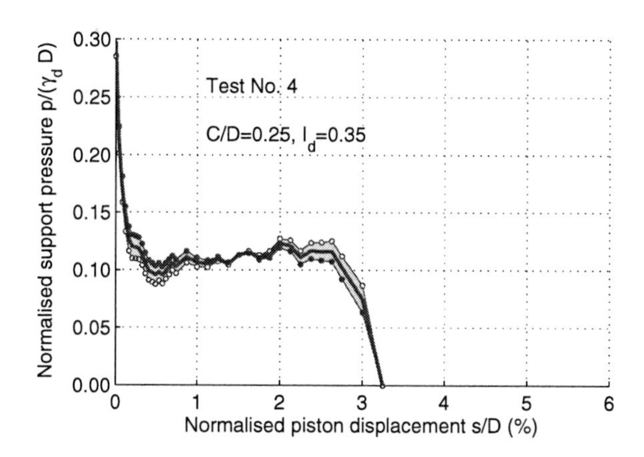

Figure B.4: Results of test No. 04

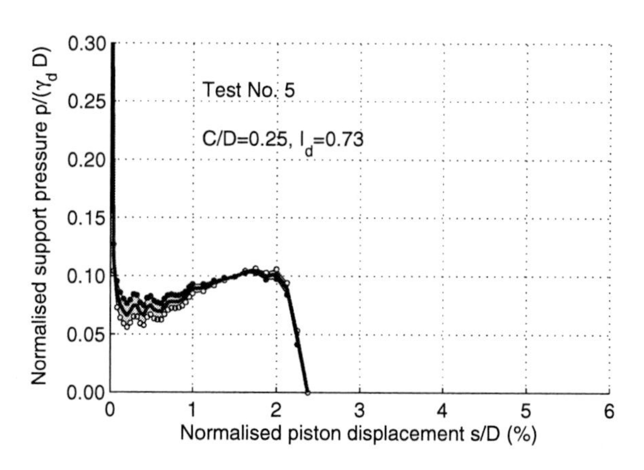

Figure B.5: Results of test No. 05

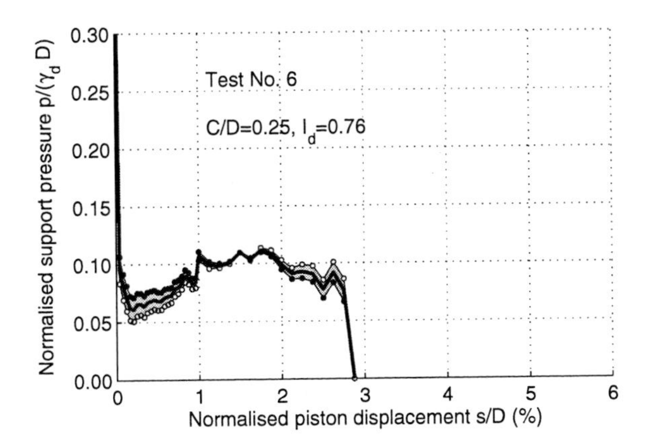

Figure B.6: Results of test No. 06

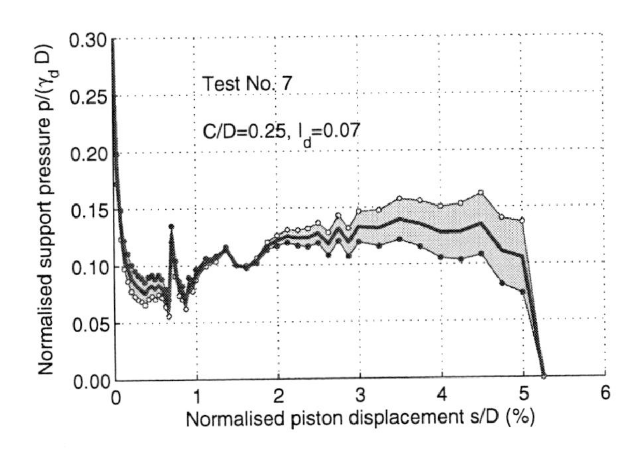

Figure B.7: Results of test No. 07

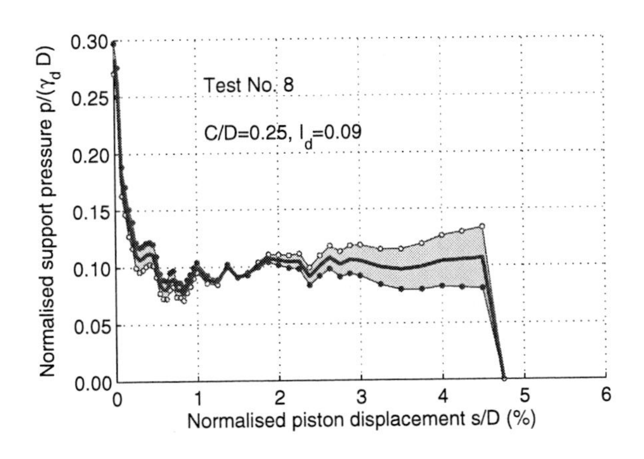

Figure B.8: Results of test No. 08

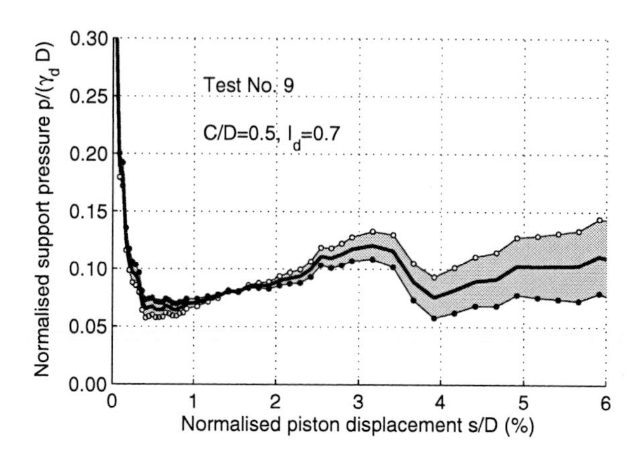

Figure B.9: Results of test No. 09

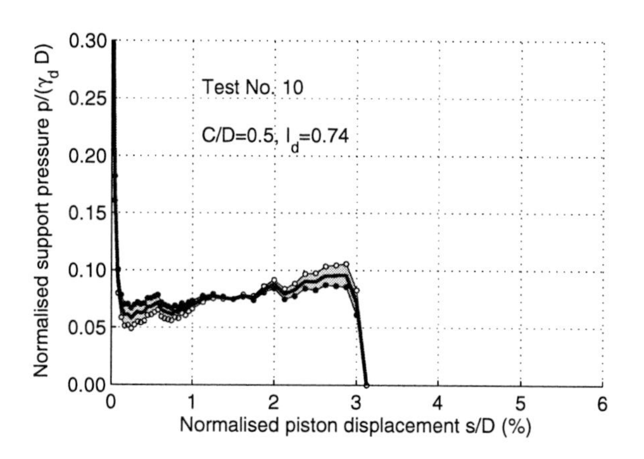

Figure B.10: Results of test No. 10

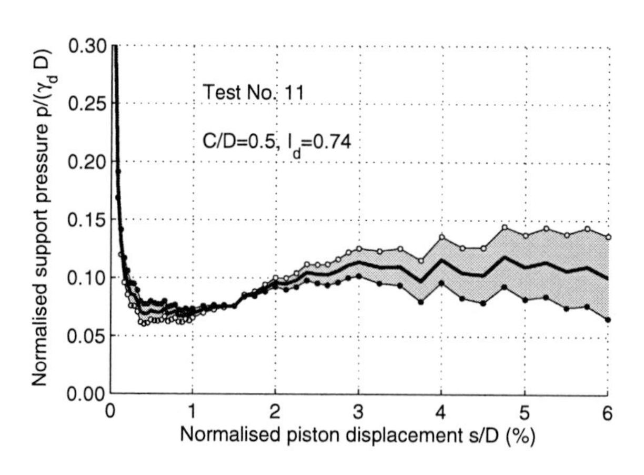

Figure B.11: Results of test No. 11

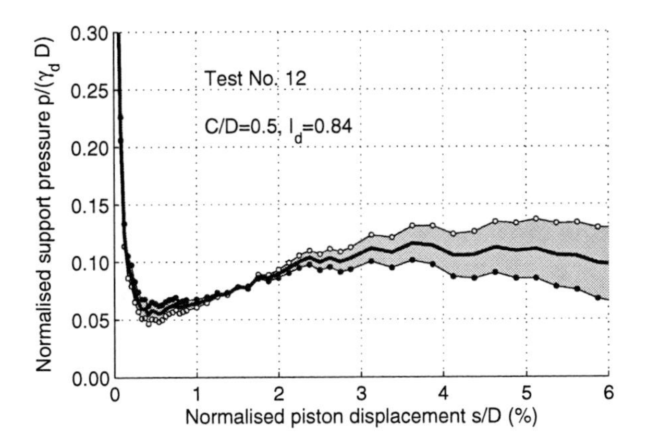

Figure B.12: Results of test No. 12

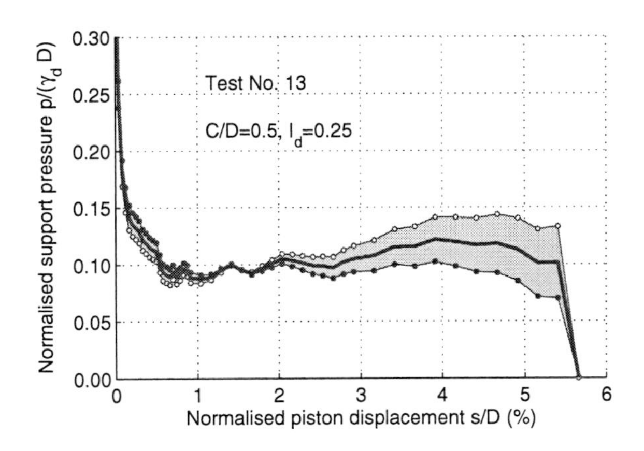

Figure B.13: Results of test No. 13

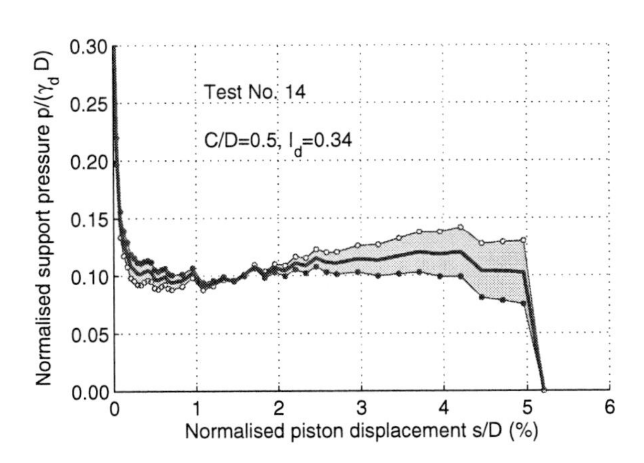

Figure B.14: Results of test No. 14

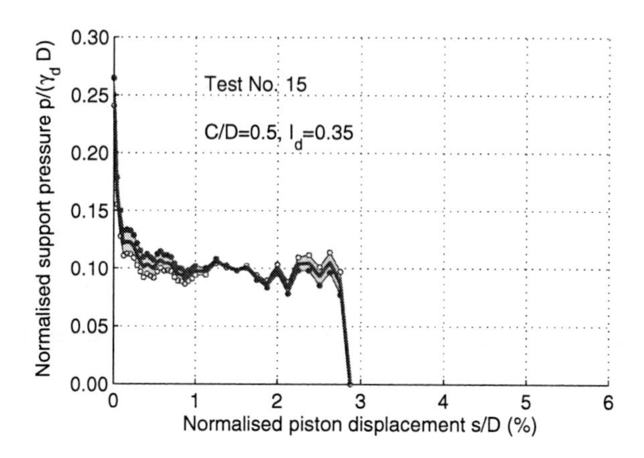

Figure B.15: Results of test No. 15

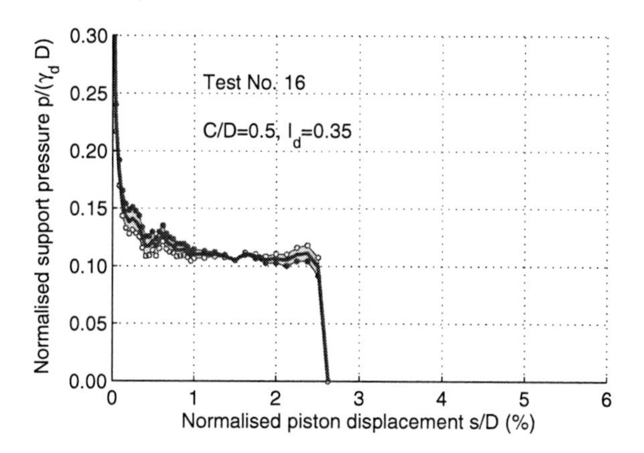

Figure B.16: Results of test No. 16

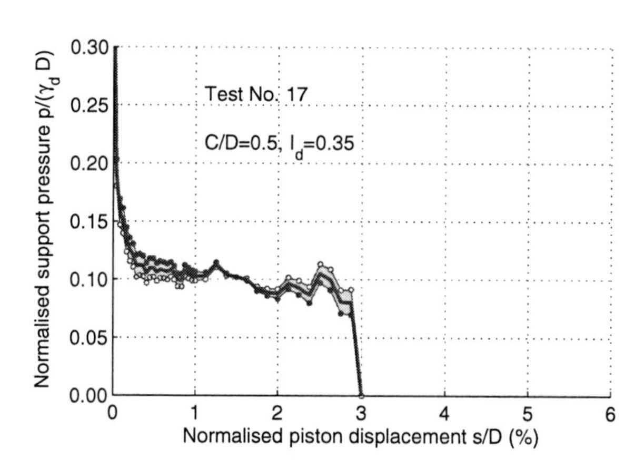

Figure B.17: Results of test No. 17

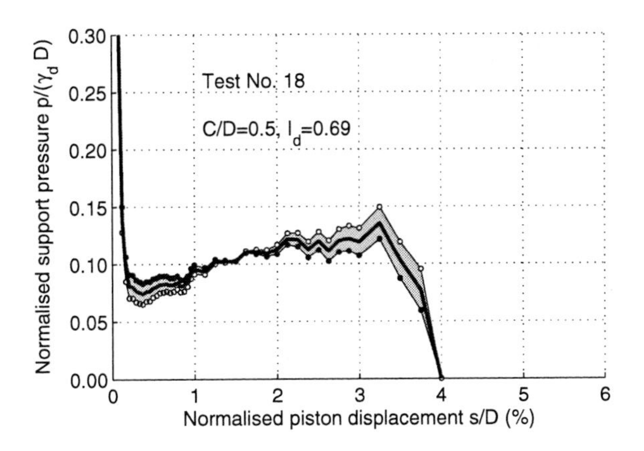

Figure B.18: Results of test No. 18

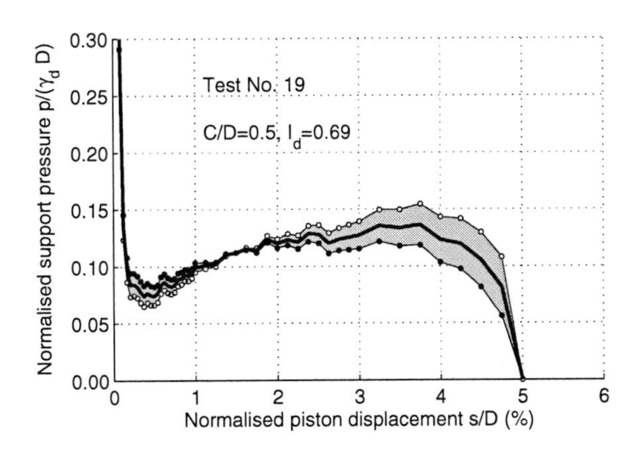

Figure B.19: Results of test No. 19

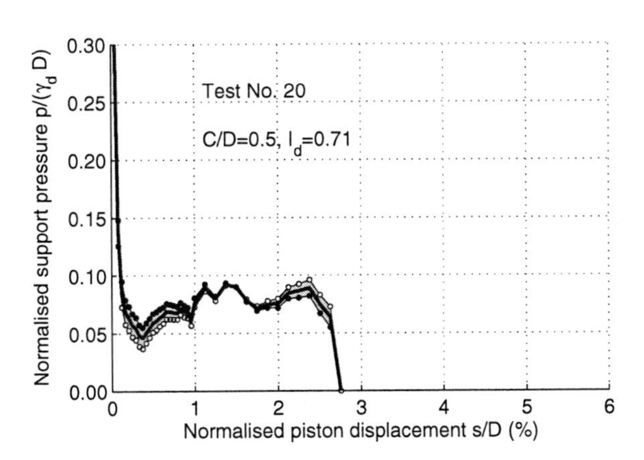

Figure B.20: Results of test No. 20

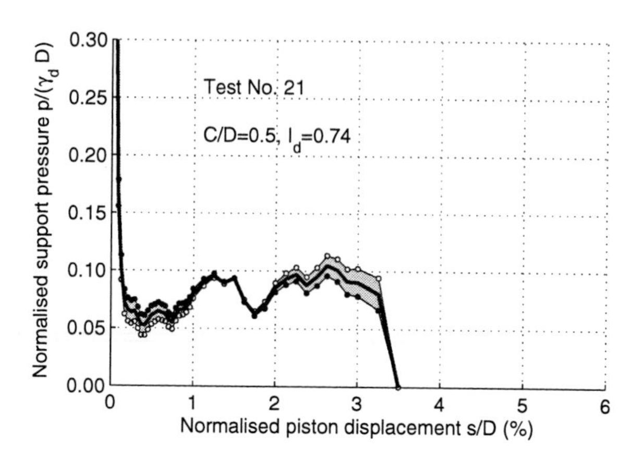

Figure B.21: Results of test No. 21

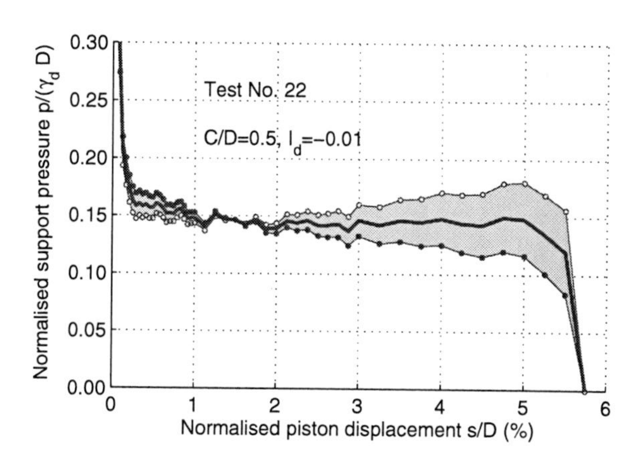

Figure B.22: Results of test No. 22

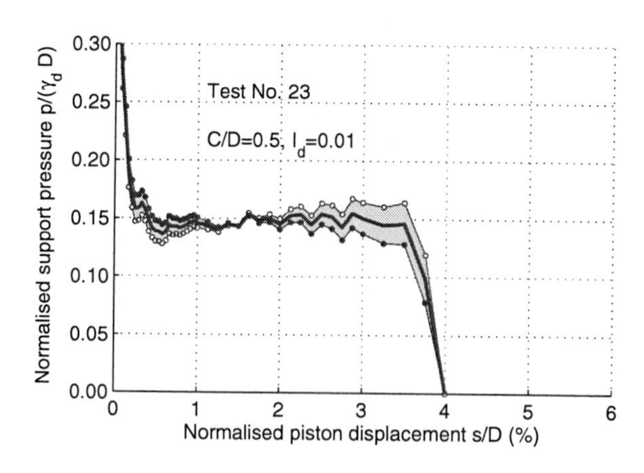

Figure B.23: Results of test No. 23

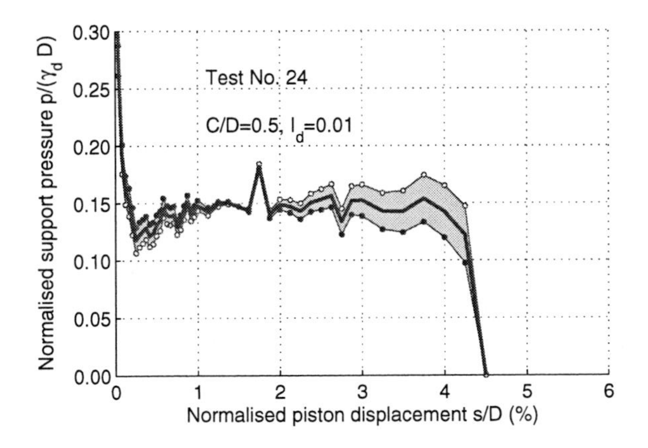

Figure B.24: Results of test No. 24

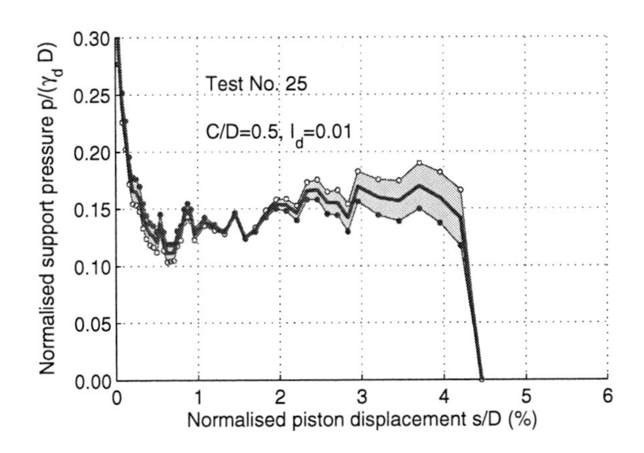

Figure B.25: Results of test No. 25

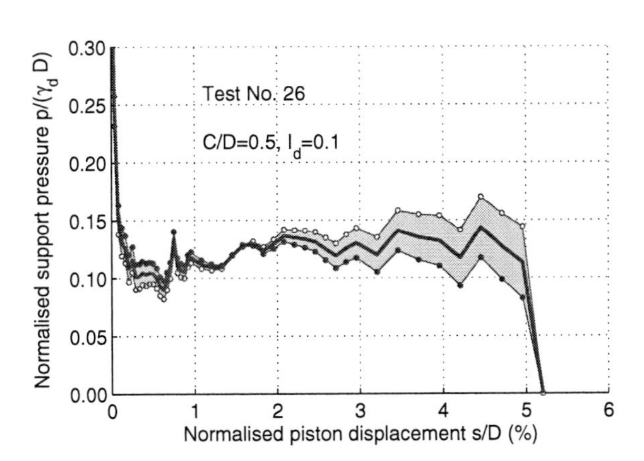

Figure B.26: Results of test No. 26

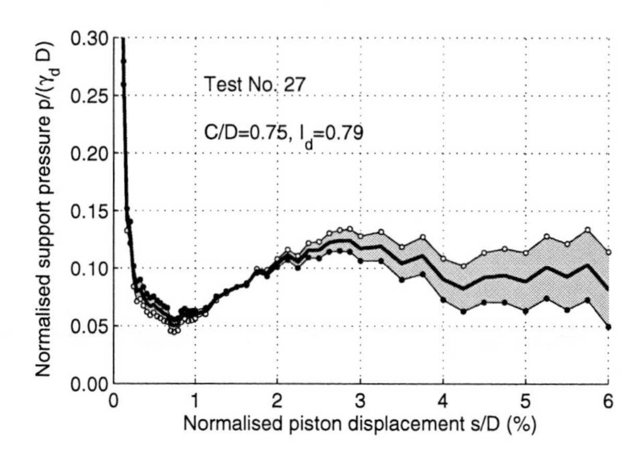

Figure B.27: Results of test No. 27

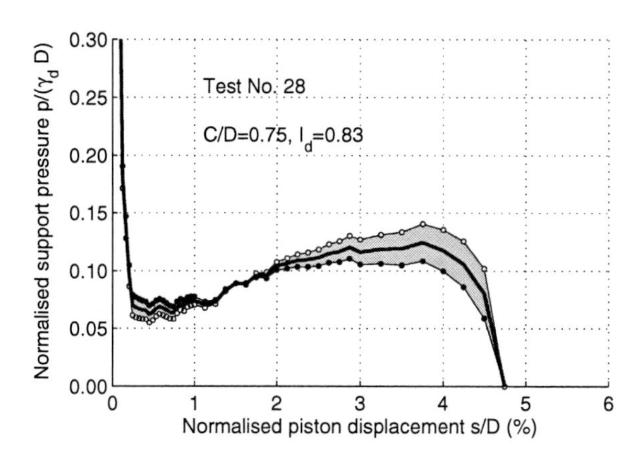

Figure B.28: Results of test No. 28

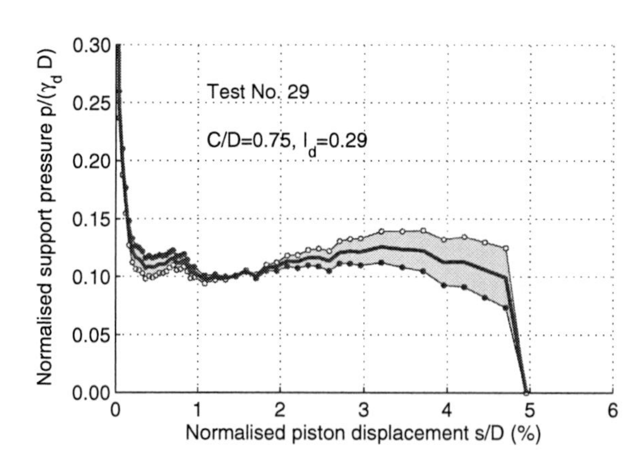

Figure B.29: Results of test No. 29

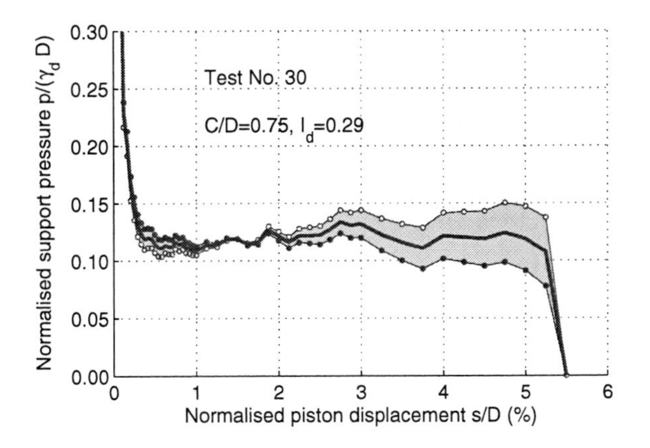

Figure B.30: Results of test No. 30

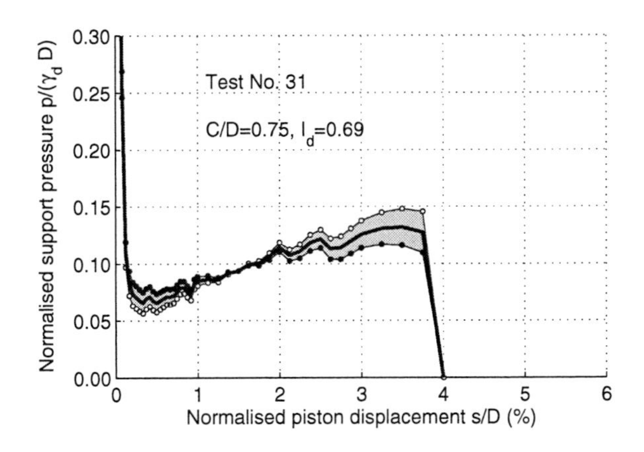

Figure B.31: Results of test No. 31

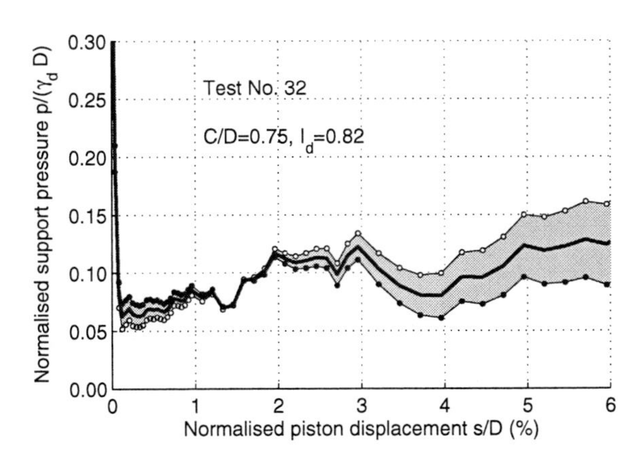

Figure B.32: Results of test No. 32

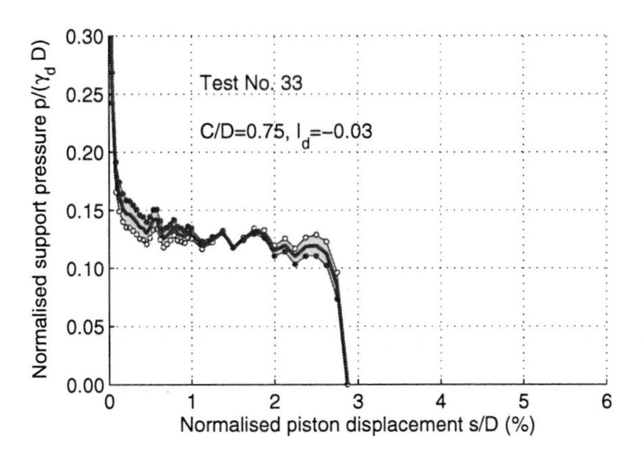

Figure B.33: Results of test No. 33

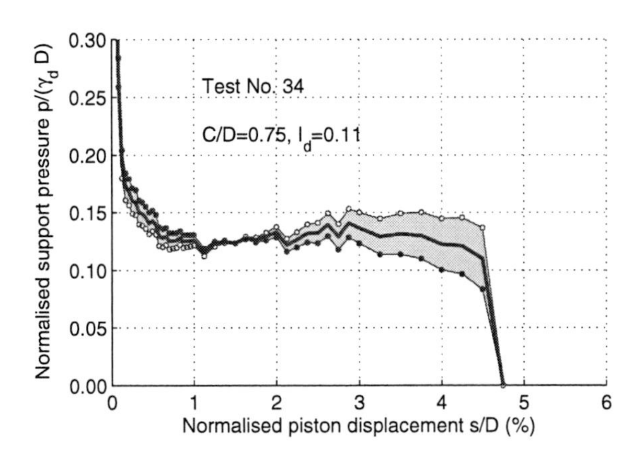

Figure B.34: Results of test No. 34

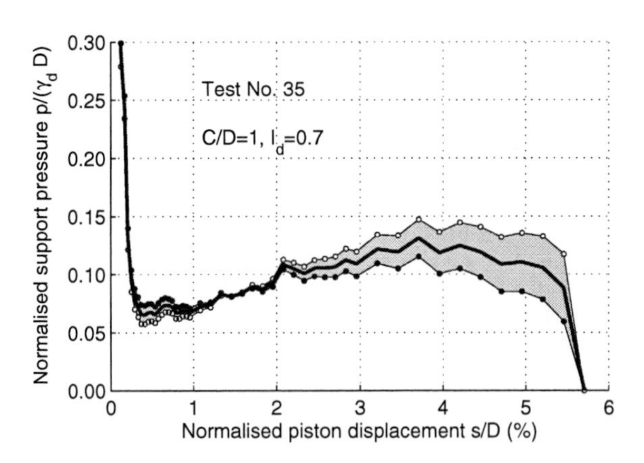

Figure B.35: Results of test No. 35

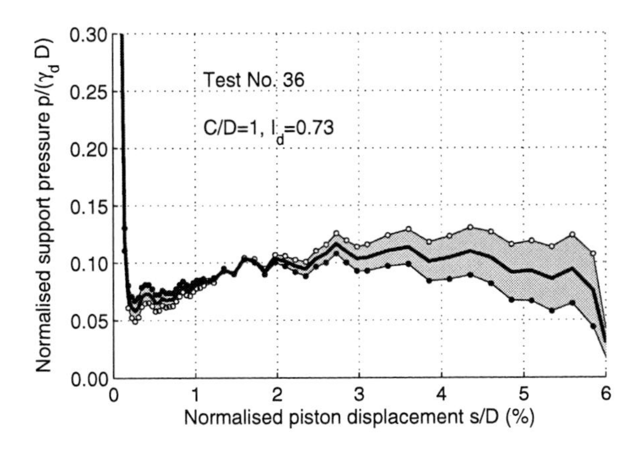

Figure B.36: Results of test No. 36

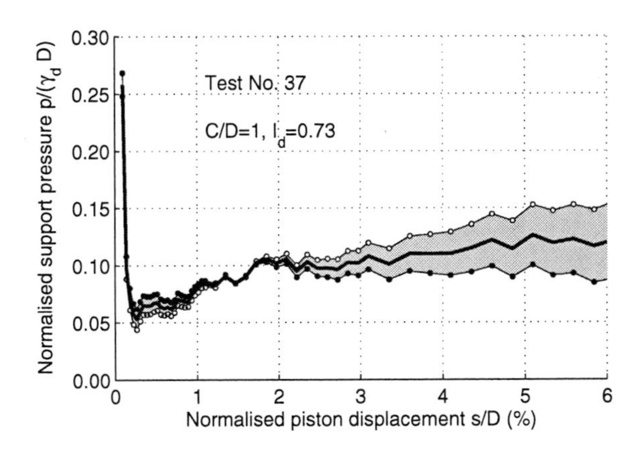

Figure B.37: Results of test No. 37

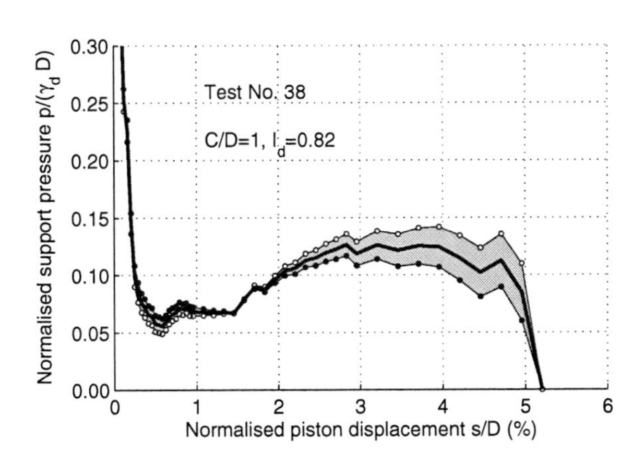

Figure B.38: Results of test No. 38

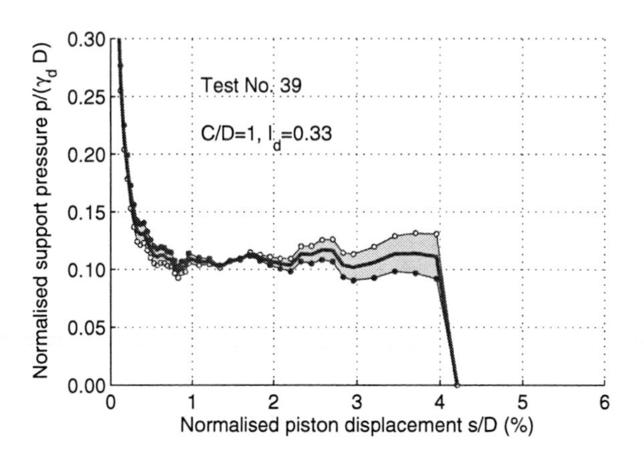

Figure B.39: Results of test No. 39

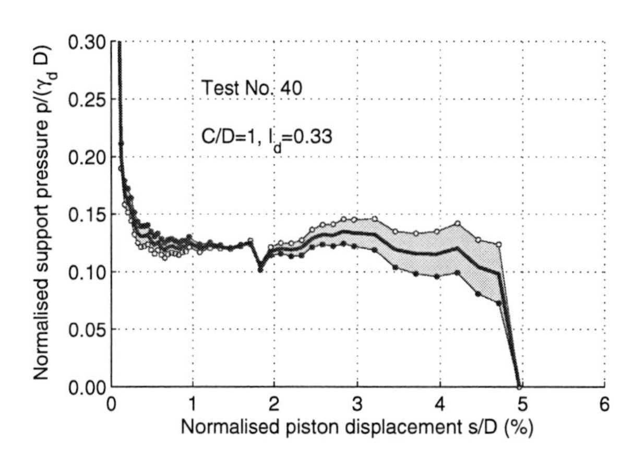

Figure B.40: Results of test No. 40

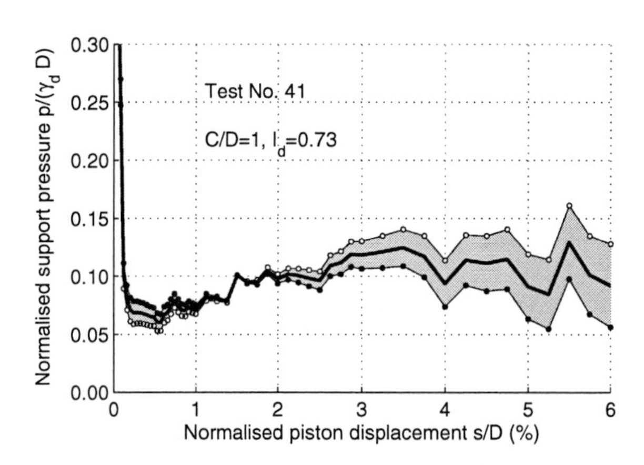

Figure B.41: Results of test No. 41

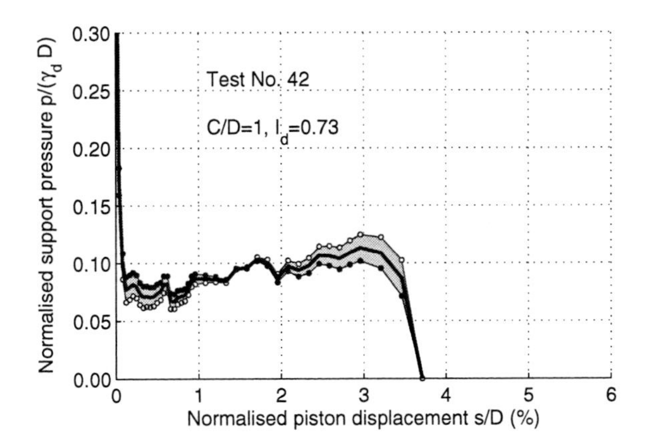

Figure B.42: Results of test No. 42

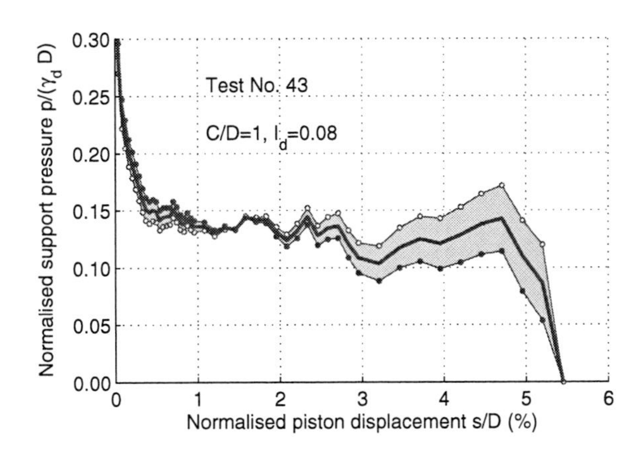

Figure B.43: Results of test No. 43

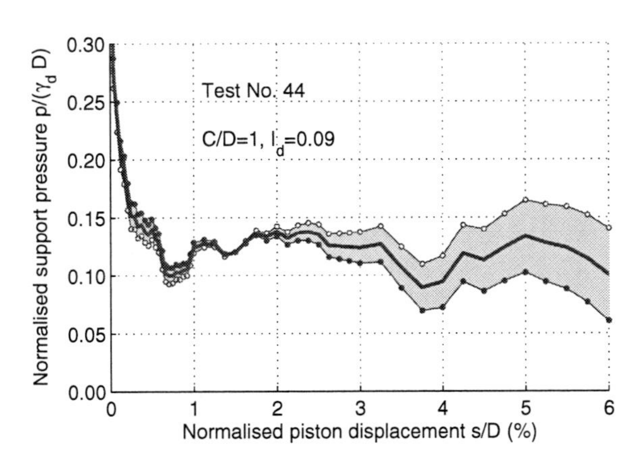

Figure B.44: Results of test No. 44

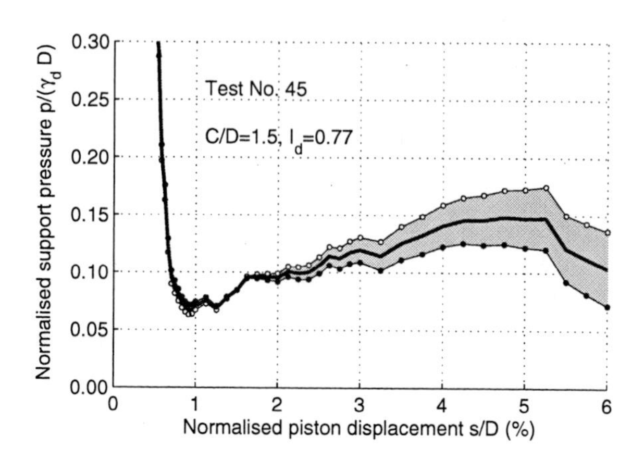

Figure B.45: Results of test No. 45

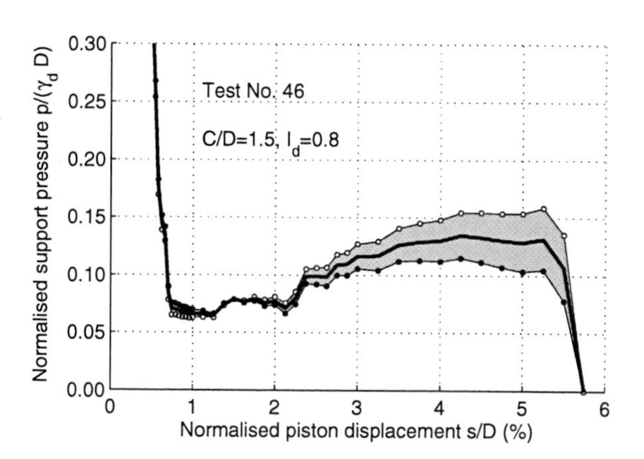

Figure B.46: Results of test No. 46

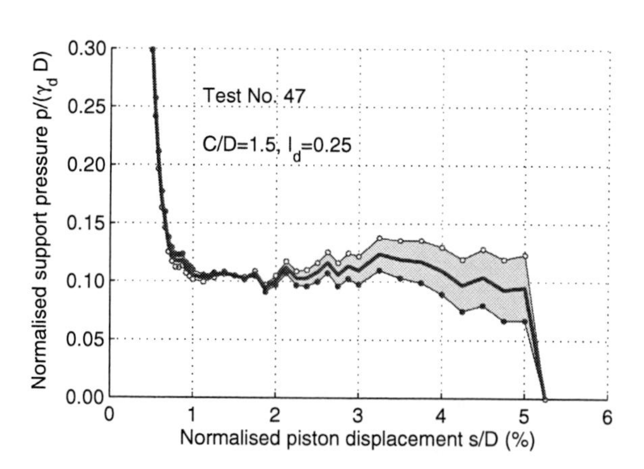

Figure B.47: Results of test No. 47

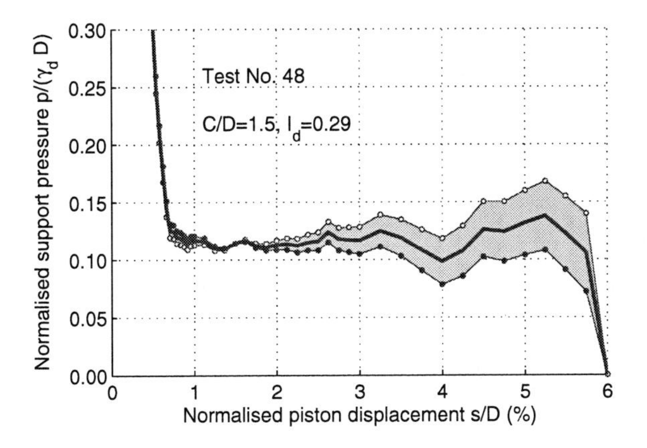

Figure B.48: Results of test No. 48

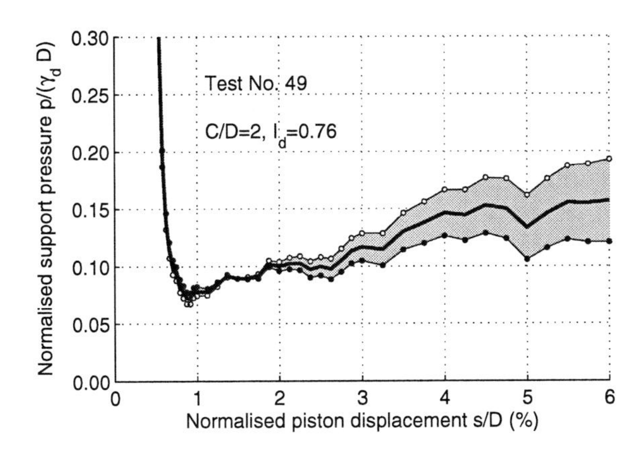

Figure B.49: Results of test No. 49

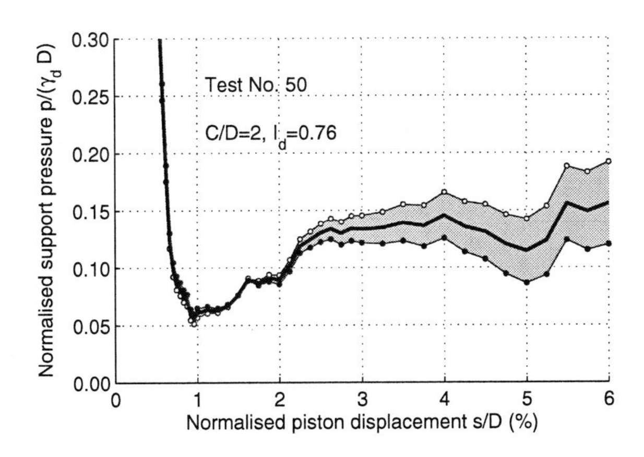

Figure B.50: Results of test No. 50

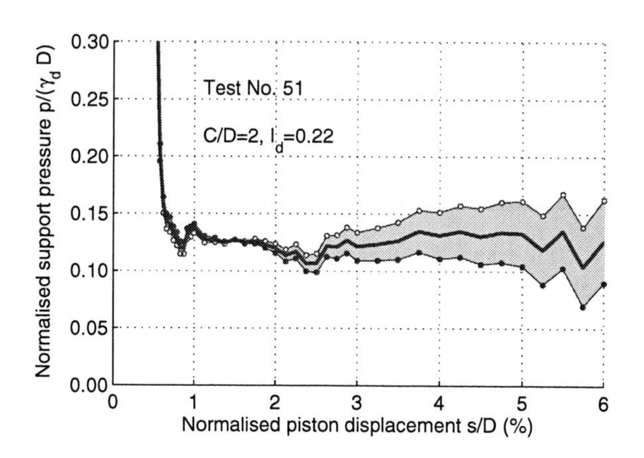

Figure B.51: Results of test No. 51

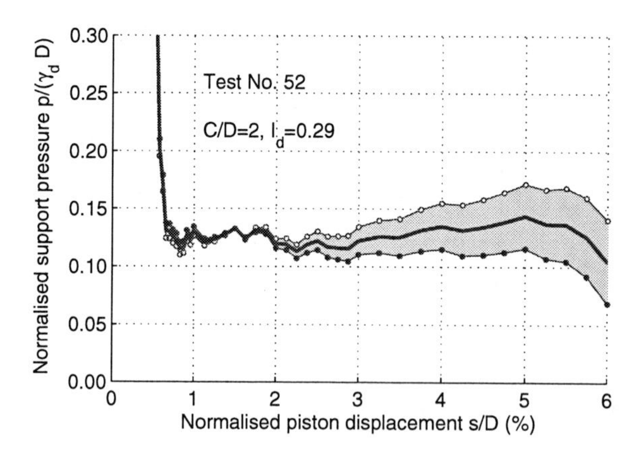

Figure B.52: Results of test No. 52

Index